THE RETURN VIEWER'S GUIDE
Important Information For Future Viewers

DON BEETON

ISBN:1537720880
ISBN-13:978-1537720883

DEDICATION

Very special thanks to my Family, my Sisters
and especially my very special Mom and Dad for all their love.
I want to thank my two children for all of their help and encouragement.
Thanks for everything! Especially all your love!!
Love Dad.

CONTENTS

ACKNOWLEDGMENTS

I have to thank my good friend Mike T. It was Mike who invited me over to look at the Moon that night using his new small backyard telescope. Thanks again buddy if it wasn't for you I never would have seen the Object. I'm very glad that we were able to catch up with each other a few times over the last few years.

My daughter surprised me with this wonderful gift. As far as I know this painting is the only painting in existence at this point in time ancient or modern that clearly shows the main elements involved in a Return. Light from outer space, the Earth, the Moon and the moon coloured Ancient Object of the Crossing Down itself complete with the Mound, the Shadow and the Standing Stone Craters.

I'm actually not a writer, that explains a lot. I have to try to write about things that should be impossible in a way that is clear so that people can start to understand the important things that I saw. My lack of writing experience makes this task difficult for me and difficult for the reader. I need readers to be patient with my writing. I'm hoping that people will remember that the bottom line for me is to try to do the right thing for my children, my family and for everyone else and for myself as well. I know that what I saw is very profoundly important. I want to thank everyone who reads my descriptions and viewing advice. I hope it turns out that you get to see the Object and you get to see in the ancient way from within the Changed and Lasting Light from the Moon.

I express my options and guesses about various topics here and there through out this book and on my website. I'm certainly not always right about things and sometimes my opinion on various questions changes as I discover and learn more. I wrote the five separate Return Viewer's Guides over ten years ago now. There have been some small improvements but basically it's in it's original form.

All of my book's content and more is available free for everyone to read and know here: www.returnviewersguide.ca

1 INTRODUCTION

I used to think that I had seen the Big Ancient Thing that everyone was looking for, now I know that I saw the Big Ancient Thing that nobody is actually really looking for.

May 24th, 1926 the big ancient place rolled down and "obscured the Moon's disc."

As a boy perhaps on the night of May 25, 1972 while looking at the Moon through a small backyard telescope I suddenly saw The Ancient Object. A speeding big forward rolling place that crosses down in front of the Moon once again in front of the Moon "obscuring the Moon's disc." My Mississauga Sighting.

A complete and total lucky fluke to be sure but I did see the big ancient place come rolling down, I saw the ancient Standing Stone Craters, and I saw the mound come forward from over the top of the Object and then transform into the overhang where the big main shadow first forms. Also I did see the sight of what the first instant in the distance of the Moon's suddenly changed light looks like and I did see the first part of the incredible visual spectacle that the big rolling place generates with the Moon's suddenly fantastically changed now incredible Lasting Light!

The crossing forward rolling speeding Ancient Object is beyond stunning and spectacular but somehow it turns out that it's all about the point of a shadow and the fantastically changed incredible spectacular light from the Moon!

Even if you decide that you are not going to read anything else if you read Return Viewer's Guide #1 you will know the basic idea of what actually happens when the ancient Object returns and crosses down in front of the Moon.

The point is people are hearing and reading,and getting a chance to hear the basic idea of what happens when the speeding orbiting ancient Object safely returns and crosses down between the Earth and the Moon.

Having heard the idea of what happens during the ancient Object's return is a very good starting point for people as they will see and know for themselves one day in the very near future.

The ancient forward rolling Object orbits down across between the Earth and the Moon. The Object intercepts and then changes the light from the Moon.

The sights I detail do occur exactly as I describe and in a repeating fashion every time the Object's shadows are cast out over and past the Object's upper right horizon into the distance of the Moon's suddenly differently behaving and differently appearing light.

Incredibly the interactions between the Moon's differently behaving light and the shadows that the ancient Object's vertically standing crater rim pieces cast into this light, happen exactly the same way every time the speeding Object rolls down across in front of the Moon from the left to the right towards the area below and beside the Sun. I do know that this is definitely true.

You can also see the exact sights that I describe. Familiarize yourself with the details and the sequence of events that I describe.

Numbered 1-5 combined the five separate Return Viewer Guides describe what happens when the Object returns and changes the light from the Moon.

Look at and follow the point of the biggest widest tallest obelisk shaped very black shadow as it races into the distance to the right. If you do this and you are able to continue looking you will see and you will know exactly what happens when the big main shadow's point suddenly sculpts and shapes the Moon's incredible changed light.

Eventually you will realize that seeing the Object and the ancient Lasting Light sights that the Object causes to happen is a goal of yours.

I am trying to put future viewers in a position were they are not at a total disadvantage concerning the sights that are about to once again happen between the Earth and the Moon.

As long as you are eventually looking to the right at the point of the big main shadow by the time the wave of the shadow is old and the building out that happens is well underway everything that you see and everything that happens just seems to happen naturally.

Believing the details I describe is not important. However having heard the idea of what happens during the ancient Object's return having read the exact details even without realizing these details are accurate people will be seeing the ancient sights and the ancient sequence of events that I describe and they will be in a situation were they will be realizing that they do indeed have foreknowledge concerning the ancient sights that they will be seeing next as they watch the next Return unfold.

Everyone not knowing the idea and the points that I describe in the Return Viewer's Guide is the alternative. That alternative is not acceptable and is not an option.

I am very fortunate to be able to describe the first part of the ancient Object's return. Later after the ancient Object once again leaves to repeat it's very long very very fantastic orbital course and the Moon's changed light returns back to it's normal state I hope it will be said that I did the right thing.

At this point as far as I know there is every reason to believe that the Object's next Return orbit down across in front of the Moon will be a safe Return just like the Object's last Return, the Return that I saw.

All the best from us,
Good luck to us all, everyone and everything.
To the future after the Ancient Object's next safe Return

2 WHEN WILL THE ANCIENT OBJECT RETURN?
May 26, 2018?

I have information that suggests to me. that I saw the Object cross down three nights before the full Moon on the evening of May 25th 1972

If this is accurate this could mean that the Object of the crossing down's orbital time period is 46 Years and 1 Day

This points directly at **May 26th, 2018** as a possible day for the return of the Object of the crossing down.

It appears that the speeding forward rolling Object crosses down three nights before the full moon

- I believe it is very possible that the celestial object that I saw was also seen and reported May 24, 1926 by German amateur astronomer Mr. W. Spill.
    - Then 46 years and 1 day later it was May 25, 1972, the night that I think I saw the Object.
    - Then 46 years and 1 day later it will be May 26, 2018

The sky was completely clear the night I saw the Object. The sky was not clear above my family home one year before I now think that I saw the Object. In Lorne Park Ontario Canada May 7 1971 there were mostly cloudy conditions starting at 11 am and then by 8 pm cloudy conditions prevailed through to clouds and rain the next day. It was not possible for me to have seen the Object three nights before the full Moon in May

1971. Overcast and wet conditions also existed leading up to, during and following the possible May 14, 1973 date for the Object's return. By 10 am May 15, 1973 the sky cleared.

This situation makes it very unlikely or more like impossible for me to have seen the Object's last return orbit down through between the Moon and the Earth in either 1971 or 1973 in the month of May, unlike on the night of May 25, 1972 when the sky was clear where I lived in Lorne Park along with basically all of south west Ontario, Canada.

I realize there are many factors and facts at work in this situation that I don't know about, don't understand or I am in error on so other than being able to state with 100% certainty that the Object I saw orbits and it will be back, I just can't say for sure exactly 100% when the Object will arrive back home down between the Earth and the Moon.

The main reason I am focused on a night three nights before the full of in the month of May is the May 24th 1926 report by Mr. W. Spill of what I believe was the same celestial object I saw.

My best guess is that the Object of the Crossing Down will return within an hour of: 9:30 pm EST May 26, 2018

Even if my best guess regarding this forward rolling celestial Object's return orbit date down across between the Moon and the Earth is not correct that would have no effect on the accuracy of the details I provide concerning what future viewers can expect to see during the first part of a Return Event.

The fully eclipsed Moon's light is suddenly seen to the right and above the crossing Object's upper right horizon. The Moon's light and light from the background of space becomes changed, and it lasts. Incredibly the fantastic ancient thing that happens, still happens and you can see it if you look up at the right time. The Changed and Lasting Light sights that a viewer sees happen repeat in exact detail.

The Object's speed is tremendous. In four or five minutes or perhaps a couple minutes more, give or take, the speeding Object travels the distance down across between the Moon and the Earth on it's orbital path down towards the area of space beside and below the sun.

Knowing the night of the Object's imminent return is all important for many reasons. My first and second book have two separate return dates that that were my best guesses at that time. Both guesses were incorrect.

This is a very difficult different situation we are all in. Again I am trying to figure out what night the ancient celestial Object I saw when I was a boy is going to roll down across between the Moon and the Earth again. When will the incredible ancient celestial Object I saw Return?

For a long time I thought I saw the Object in 1970. Now it turns out that this was not the case. Instead it does look like 1972 was very probably the year I saw the Object that I describe on this website. I really wish I could have realized this sooner.

There is a long path that has led me to the May 26th, 2018 date.
This does not mean I am correct this time and that the May 26, 2018 Return date for the Object is guaranteed.

I can guarantee that the celestial Object I describe on this website will in fact return and change the light from the Moon exactly as I describe. Sooner or later the Standing Stone Crater covered Object will be back that's 100% guaranteed. It does turn out that there is a possibility that the incredibly massive incredibly ancient forward rolling speeding celestial Object I saw crossing down while rolling towards the Earth from it's crossing position in front of the eclipsed Moon actually might arrive down once again finally back home from the depths of space on the evening of the twenty sixth of May 2018.

3 THE RETURN VIEWER'S GUIDE

There are five separate Return Viewer Guides. Each version describes the ancient celestial Object that I saw, and the spectacular very special visual event the returning orbiting Ancient Object's forces causes to happen to the light from the Moon and the light from the background of space. Each version adds more, the same and or different points. Also I include thoughts and comments. I hope that all of the separate versions, added together, will give future viewer's a fairly clear picture of exactly what they can expect to see and the important timing of the order of events during the Ancient Object's next return orbit down across in front of the Moon.

Because I saw the Ancient Object through a small backyard telescope, I am only able to provide viewer tips from the extremely up close telescope aided perspective that I had. I have no idea how far away the details I describe would look to a viewer who is viewing with the naked eye.

A decent modest, small backyard telescope is all that is required for an extremely up close fantastic view of the returning crossing Ancient Object, and the incredible visual spectacle that it generates with, in, and then out of the lasting and then eventually moving changed light from the Moon.

Return Viewer's Guide #1 is in point form and it is a good place to start. Even if you decide that you are not going to read anything else if you read Return Guide #1 you will know the basic idea of what actually happens when the Object returns and crosses down in front of the Moon.

A basic point that I have to make is that I don't have any background whatsoever in the area of telescopes. I only know one thing and that is what my view through the telescope looked like that night during the Object's last return.

In the Return Viewer's Guide I describe landmarks and the Object's surface features and the overall scene also most importantly specifically what to look for and when and were to put your eye and what incredible sight will be seen there! Also more incredible sights are suddenly seen in the Moon's very mysterious fantastically suddenly changed light. Later, after the first instant all of these sights and the focal point of the Changed Light the Man of Light The Awesome Good and the other main incredible sights and specifically other human shaped main forms that are aligned in a very specific fashion oriented around and above the object's upper right horizon during the time when the Changed Light is rising. I describe this very specific double row's curving staggered alignment and many other specific details that will help future viewers of this incredible event. Also I describe the object's flight path and it's motion and specific details concerning the order in which the sights and events occur that I am familiar with on the surface of the object that I viewed completely by chance through a small telescope in approximately 1972.

My viewer's guide is just a small starting point for people because once their eye's gaze reach's the ancient place in the Moon's incredible changed light at that instant in time, at least every telescope aided viewer is going to be suddenly feeling way way more than just completely totally alone and overwhelmed by this staggering penetrating sight, that is simply completely beyond belief or even any sort of comprehension!

It's really there and it still happens!
It's the mind bending sight of the Awesome Good!
I saw it and then I saw what it does next, and then I saw what it does after that along with the rest of the impossible sights that are also suddenly there! Defiantly this is way more than just every human for himself although that does describe it! Clearly the impossible is way more than just simply there to be seen. If you see this incredible sight, instantly you are literally even physically forced to comprehend it!

Now you have seen, and gone face to face, with the all time greatest ancient human mystery at the same moment it seems as if suddenly you are realizing that it turns out that you practically really don't know anything at all! At least that was the effect for me. How could this be? When you look up you are simply not going to see some things, most things. When you look up, you are simply not going to see anything familiar except for those big space things that you would expect to see when you look up into space at least that's what I always thought.

It turns out that things are not as simple as somehow a thing, that should be impossible, does happen. Not only that but to see is to know that not only does the impossible happen, but the impossible repeats itself and it actually somehow happens many times over! You would think that seeing the ultimate ancient mystery would provide you with some sort of, even all knowing insight. Defiantly that's not the case. Seeing is what it's all about. Completely knowing everything about the big how and why questions, is surely never going to happen for anyone, ever.

Well short of any sort of complete understanding I can detail this sight. I can also point to this exact look right here, down on the Earth. The ancient people carved in stone. The exact specific look is here and I can point it out. It turns out that this look is well known and is already a great ancient Earth mystery! Even just this one specific area of things is a truly fantastic thing to know!

Although that look, is down, here, it's not as simple as that. The way the Awesome Good appears, changes. In the first instant, you see it from a certain position and a certain angle, from close above, and behind. Then, it turns around! Suddenly the intensity focuses directly on you through the Awesome Good, the focal point and heart of the lasting light event.

The Moon's Changed and Lasting Light looks like a place up there to the right of the Object first a valley then a mountain that rises growing at it's base and in between the human form is created by a shadow in and out of the changed light itself. The muscular Man of Light is a solid looking sight that is light that has been shaped and formed by the point of the big main obelisk shaped shadow in the Moon's changed and lasting light! It's somehow real and over there as if a statue. A statue is something I can compare the focal point of the lasting light event with but it's a lot more than just a statue but for some sort of comparison it's at least a start.

Without any question the heart of the Changed Light event is the Man of Light. When you first see him he appears to look a certain way. This first look that he has is the way he appears when you first see him and then he changes and evolves!

The face of the Awesome Good seems as if it retains it's central characteristics but the way that he looks outwards from the edges of his face do change. This causes the overall appearance of the Awesome Good to change and evolve while continuing to retain the same basic face. It's not as simple as saying the face of the Man of Light stays the same because there are some slight changes there as well. I try to describe this in more detail.

During the rising light mountain phase of the lasting light event the Awesome Good the focal point of the lasting light event maintains his second new look. I can guess that there is a third evolution that follows an evolution that I didn't see but I think that the third next specific look is seen on Earth carved in stone. An incredible thing! The main central way the Awesome Good appears or looks remains but it's overall look does change. I saw the first two looks up there in space over and then above the object's upper right horizon and I can see it's third evolution down here carved out of stone!

The basic idea is to find the big main obelisk shaped shadow that emerges from the shadow's awesome wave as it races away from the viewer at an ever increasing angle towards and into the distance to the right over,and later off of the object's upper right horizon into the distance of the incredibly changed fantastic light from the Moon.
Find the very point or the top of the big main shadow and focus on the idea of continuing to remain looking at the big main shadow's rapidly moving point as it travels away for you towards the object's upper right horizon.

If you are able to do this then your eye will literally seem to arrive drawn forward to what somehow must be the all time most incredible ancient place, at the all time most ancient and incredible instant in time just at the exact instant that the back of the incredible Awesome Good is seen right there suddenly formed! Then the Awesome Good does the most incredible thing! He turns around to his left and at the same time he gets up and looks up directly back strait at you as if he somehow noticed and knew that you were there looking before he turned and got up! I call this The Big Noticing.

The impossible is really there and very suddenly it's as though it's noticed you! Suddenly the light's incomprehensible mind bending pressure and intensity is on! At this point the full force and intense pressure of the moving lasting light is yet to to arrive but at the very least you will be realizing that the sudden surprise and shock that you are feeling because of what you will be seeing at this point is causing you to comprehend that you are really seeing the ancient sights in the ancient way. The Moon's changed light now focuses on you directly right through the sight of the eyes of the Awesome Good if you can remain looking!

Incredible seconds to know. These seconds can be known by anyone who looks. The idea is that not only is there a returning object but there is an incredible sight that once again will be seen there that is generated by the object and sculpted and created by the point of a shadow in and then out of the totally incredible changed and solid and lasting and moving very magical light from the Moon. It turns out that this is just simply the way it is and basically by a very lucky fluke I saw this happen.

The incredible sight of the Awesome Good must surely be the all time most ancient and fantastic ultimate human mystery! Also surly it's our greatest most precious and fragile treasure! This is a sight that will be there to be seen if you know to look and if you know were to look.
It's a natural. Find and let the point of the big main shadow lead you and you will arrive and then you will see the big sight. It turns out that there is a Big Thing and this is it!

I will be trying to describe very specific details that will describe the sights and land marks and the timing of the order of events that are seen that lead up to the incredible sight of the Awesome Good. Also I describe what happens next.

A short transition time period exists. During these few seconds the Moon's and expanded and spread out light pulses inwards. The outer edges of everything that is the Moon's changed light shrinks inwards. The Moon's spread out lasting light's expanded outer circumference retains it's outer outline and overall shape and inner details and becomes an inner circumference or inner area ultimately a sunken area down within the upper structure of the rising light mountain as I try to describe it.

The Moon's light is somehow changed by the forward rolling slowly sinking object's forces. After the shadows that are cast off the object interact with the Moon's changed light the changed Moon light somehow lasts. The Moon's light is changed then it lasts. The shadows sculpt and shape the lasting Moon light. Everything that you already saw happen

11

remains or is changed in some manner. Now an intense pressure and direct focus arrives.

All of the inward pulsing has ceased and suddenly the lasting light is growing only at it's base and it is bending, curving moving directly towards you. The solid looking, lasting light contains all of the impossible incredible sight's down within the top area of the light's rapidly rising upper structure and it is getting rapidly closer. Now you will be feeling the full pressure and you will be realizing that the pressure that I describe is on. Not only is the pressure on but you will be very startled knowing that you are literally experiencing sensation a new or unknown feeling and sensation. This is definitely true at least as far as telescope aided viewers are concerned.

Everything that is happening and everything you are feeling and all of the intense pressure and everything that is indescribable is now focused very specifically into you through the heart of the lasting light. The heart or focal point of the lasting light is definitely the sight of the Awesome Good!

I think that it is also worthwhile for me to include comments that describe and give some some background information. I hope that this may help to describe my extremely limited and narrow viewing perspective and my viewing situation as a young boy in approximately 1972.

I can describe in specific detail the sights and the order of events however this is a general viewer's guide. It seems to me that when a wide array of many different sized backyard telescopes, and many different levels or powers, of magnification are taken into consideration, all I can really hope to do is to provide a general guide for viewer's.

I don't know telescopes and because of this I don't understand the size, magnification and width, and depth of field issues and how these issues will effect the location of the viewing position that future viewer's will find their eyes seeing from or the width or size of the picture that they we see. This gets into hard areas for me to quickly explain. I know what my view looked like and I know how the place my eye was seeing from seemed to race forward and then race back to what you would think of as a more normal seeing from back were you would expect, viewing position.

Whenever you look back left against the movement of the shadow's wave or especially when you focus on the point of the big main shadow, at some point you find yourself moving closer. An incredible effect that will be noticed by future viewers especially when you arrive at the ultimate instant in time at the ultimate place in the light.

At that ultimate instant when the Awesome Good turns around you are at your most furthest forward viewing position. The sudden incredible face to face lasts mere seconds perhaps a couple but this point in time does last! Then as I describe the very sudden shrinking down of the outer perimeter of the view and the transition period between phase one and phase two of the illusion starts. For split seconds you watch the light shrinking pulsing inwards and the focal point wiggling around and changing.

Then suddenly you've moved back and at that point for me I really couldn't get back fast enough. At first your eye is eventually drawn forward and then your eye races back. An incredible effect that makes explaining exactly what people are going to see and how people are going to see it extremely difficult when I consider that vast array of different types of telescopes that I really actually know nothing about.

The telescope size, magnification, power and depth of field issues leave me convinced that all I can do is try to describe how things looked to me. The telescope that we looked through that night was a new small backyard telescope. I know that for me personally next time I certainly don't what to look through some sort of substantially bigger telescope than I did last time. I understand that I will see from close enough, at least for me with just a small modest backyard type telescope.

I very sincerely feel as though some sort of larger telescope especially a telescope that is a lot larger would be way to much telescope at least for me. It's not as simple as the face to face par, is more than just close enough for me it gets into questions like how much of an overview of the entire object and changed light are you going to see with that extra powerful telescope and a smaller field of view?

I understand that it is possible to see past the edges of the object and most of the Changed Light while still being able to easily get right up close at that big moment when the Awesome Good turns around using just a small modest telescope. What more is there?

I wrote all of these separate short return viewer guides at separate times and usually separate from each other. Once again there is repeated material. I thought that instead of trying to change this and change that, so that there was less repeated material I would just leave everything the way it was as not to take away from any one version.

I touch on different topics and try not to explain these separate topics and points in lots of detail. In stead I have tried to streamline and simplify the separate viewer guides by using other pages to try to explain more about such things as more overall background information, the phases of the changed light event, the shadow, the far slope, the large and the small shape effects etc and the Awesome Good.

Because of my lack of experience as a writer a lot of patience is always required from you the reader. I would like to thank people for their understanding and their patience.

As always I try to remain focused on the obvious best part all of the good, the Moon's light and especially the obviously incredible Awesome Good the Man of Light the heart of the changed light from the Moon.

Return preparation issues are the flip side for me. I have to say what I think and I know that it's very important to go there. However it is good to know that the good part is really good! I am always in search of ancient texts and material that explain to me about what happens next up there in the light after my view of it ended. I know that these things I read are as real as the lasting light and the lasting light is real light from the Moon that does somehow last.

People can be sure that there is a measure of real hope. There are many many troubling things that are written in the past that are clearly discussing the way the Moon's light looks and behaves! This gives me great hope that my distress may be at least in part for not! I feel better! I know and understand that this understanding that I have may not translate into complete safety on the ground down here on Earth. I sure hope it does though. I just don't know. It is important that people realize that they can know that hope does exist.

Once the object is sighted the distress and terror will begin. Hope can also start as well. I also know that the very ancient Awesome Good is returning! The Big Noticing, will happen!

Our ancestors knew these incredible awesome ancient moments and were always looking forward to it! I try to use this as a guide for how I should feel about it. Now it is our turn to join with our ancestors! Finally we can all also see the Object of the Crossing down and profound sight of the Man of Light! We can also see and know at least in part were we come from! We are conscious living matter copies of a shape that is light!

 I sincerely hope that I will be able to help a great number of people to also see the ancient Lasting Light sights. Once again, there will be many human eyes riding the point of the big main obelisk shaped shadow, together into the distance of the Moon's incredible light. Very soon we will all meet at that indescribably special place in the light and we will all be suddenly noticed and then seen by the incredible sight of the Awesome Good!
This is going to be very special!

If you are reading this I want to say that I hope that it will turn out that I will have been able to help you have a chance to see and I hope you do see and I hope that your whole family also has a chance to see especially your children!

4 RETURN VIEWER'S GUIDE #1

1. The forward rolling moon coloured object rolls down from the area of space above and to the left of the Moon.

2. The object covers the Moon's upper left area first before it completely obscures the Moon.

3. As the object rolls forward a large raised surface feature comes into view from over the top of the object.

4. The Moon is suddenly seen to the right of the object.

5. The Moon no longer looks like the Moon at this point, instead the Moon looks like a fantastic tremendous valley.

6. As the object continues to roll forward the raised area becomes a smooth curving steep cliff.

7. The raised area then gradually develops into a very large smooth and rounded overhang.

8. Suddenly a shadow that looks like a big black spot appears under the overhang.

9. The black spot of the shadow spreads out and thickens under the continuously growing developing overhang.

10. Suddenly a certain critical point is reached and the shadow suddenly leaps down to the ground below.

11. Immediately the wave of the shadow races to the right and travels away from the viewer at an ever increasing angle.

12. The shadow is going to pass over and eventually travel off the object's surface to the upper right into the Moon's light.

13. A series or group of tall narrow individual shadows emerge and travel further and faster than the shadow's main wave.

14. At one point this series of tall shadows look a lot like a line of telephone poles stretched out across the countryside.

15. The group of tall individual shadows that are seen are spread fairly evenly from left to right beyond the shadow's wave.

16. A group of three main obelisk shaped shadows emerge and travel further, faster than the rest of the main shadows.

17. In between the two outer smaller obelisk shaped shadows a central larger wider obelisk shaped shadow is seen.

18. The big main obelisk shaped shadow is twice as wide as the two smaller obelisk shadows that trail one on either side.

19. This central obelisk shaped shadow is the big main shadow that ultimately sculpts the form of the Awesome Good!

20. The central and most important main idea is this; Find and focus on the point of the big main shadow as it races away to the upper right.

21. To the left if you see normal surface area past and over the raised area, you have only seconds left! Look back right!

22. Once again, this is important! To the left meaning if you are looking at the hard ground of the object itself.

23. To the left if you can see beyond and over the raised area and you can see normal surface area look to the right!

24. Look back to the right right away because there are only seconds left before it's time to get noticed!

25. To the right locate the point of the big main black shadow and start to focus on the idea of not looking away!

26. Find the big main shadow positioned centrally between two smaller shadows that accompany but trail on either side.

27. The main now very tall obelisk shaped shadow is now traveling very rapidly into the distance of the Moon's light.

28. Suddenly the big main black shadow's point starts to go up and down and up and down very rapidly.

29. The ultimate moment is about to arrive. You are about to see the sight of the muscular Man of Light from above and behind.

30. Instantly after the up and down and up and down stops the shadow's point starts to cut to the left and to the right.

31. The shadow's sudden zig zagging squiggly line is now sculpting and forming and creating the Awesome Good!

32. Suddenly the race forward has stopped and the big main obelisk shaped shadow is gone.

33. Now not moving forward or backwards you look down on a peaceful and tranquil fantastic scene!

34. Suddenly in extreme detail you clearly see the Awesome Good's very muscular back!

35. The view down from above and behind lasts just long enough for you to be able to realize what you are seeing.

36. The sight of the Awesome Good gently drifts up closer towards were you will find yourself seeing down from.

37. Also, a flat area is suddenly seen around were the Awesome Good is positioned.

38. Past the outer edges of the flat area the misty white indiscernible area is still visible. I try to discuss this in detail.

39. Each successive newly arriving one way wave pulse of rising lasting light builds out then down.

40. Another instant later it happens! What must surely be one of the all time most incredible sights in history!

41. The incredible Awesome Good turns around to it's left and gets up to squarely face you all in one very fast motion!

42. The Big Noticing! The sight of what should be impossible has just spun around and looked you strait in the eye!

43. This ultimate very up close first moment lasts only long enough to go face to face for only a very few seconds!

44. In that time he drifts up twice to his closest point, then seeing from farther back you remain face to face while rising.

45. The first part or phase of the Changed and Lasting Light event, the Valley is about to end.

46. Between phase one, the Valley and phase two, the Light Mountain, a short transition time period exists.

47. The ultimate staring contest lasts from the big noticing through the transition period, into the rising Mountain phase.

48. The transition period between the valley phase and the mountain phase, only lasts for a few seconds.

49. The outer edges or outer perimeter of the Moon's light, everything that was the valley shrinks, or focuses inwards.

50. This is the same point that the position that you are seeing from rushes back away from the Moon's changed light.

51. During the transition a shrinking pulsing inward effect is repeated and briefly seen, a number of times.

DON BEETON

52. At this point the Awesome Good wiggles around, reorienting itself moving very much like a cat about to pounce.

53. This is also when the way the Awesome Good looks or appears overall changes and evolves into it's second look.

54. This is also the point when the rest of the shadows that trailed the main shadow are now sculpting their own figures.

55. This is also the point when the Moon's light seems as if it is finally arriving from over the object's upper right horizon.

56. Instantly after the inward pulsing effect is over the full force of the pressure and intensity of the Moon's light is felt!

57. Simultaneously an incredibly startling sight is seen as the entire area of the Moon's now solid lasting light, rises!

58. Everything that was the valley, the Awesome Good included is reduced in size and is located down within the top!

59. This completely incredible area of intense solid lasting light grows at it base with every newly arrived area of light.

60. The staring contest and the situation for the viewer grows more complicated as the Awesome Good nears!

61. An impossible indescribable situation that somehow really is happening is now rapidly closing on your position!

62. Real light from the real rock of the Moon makes this growing bending column of rock look very real.

63. Phase two. A rapidly rising bending curvingly strait rock mountain travels toward you right up through space!

64. Down within the rock mountain's upper structure what should be completely impossible is seen!

65. The Awesome Good still stares at you but it has evolved and has it's new second look or appearance.

20

66. The other main shadows have also sculpted familiar forms. All of these forms are similar and familiar looking.

67. The valley is still there down within the very top of rising column of the lasting and very incredible Moon light.

68. The valley is now seen from a higher angle during the mountain phase when compared to the earlier valley phase.

69. The valley now looks like a sunken room. The flat area first seen around the Awesome Good is the floor of the room.

70. I attempt to describe the sights contained down within the mountain's upper structure in more detail elsewhere.

71. I also attempt to provide more details concerning what the overall scene looks like at this point as well.

72. Describing the light's pressure and intensity is virtually impossible. The intensity and pressure are very strongly felt!

73. Seeing is what it's all about. Once you see, it is very difficult. Keep looking! Stay looking! Don't look away!

74. The actual object is rolling much lower in the scene at this point as it continues moving across to the right.

75. Everything you see that is not the object, but instead is light, is oriented and focused on and moving towards you!

76. The object is moving basically in your direction and the Moon's light's light is definitely moving in your direction.

77. At this point my view of this event was about to end. I describe and detail how the staring contest ended for me.

78. Below I expand my return viewer guides in an effort to further describe the various sights that are seen.

79. I hope that all of my efforts taken together will help future viewers see sight of the Awesome Good and beyond.

A vertically standing crater rim piece positioned on the crossing Object's surface casts a shadow into the Changed Light from the Moon. The shadow's interactions with the Changed Light cause the light to become shaped into the muscular form of a Man. This shape then lasts as if a statue.

This impossibly random series of events is set off in motion by the Ancient Object eclipsing the Moon and it's powerful forces then causing the Moon's newly emerging light to look and behave very differently. Starting with a randomly standing crater rim piece casting a shadow into the great distance of the Moon's Changed Light.

A face and the very muscular form of a man that are solid Lasting Light from the Moon, both sculpted, shaped and formed and created by the point very long tall black shadow.

The Awesome Good is the Man of Light and the mysterious way the Changed Light imprints itself into everyone who looks. A part of what I think of as the heart of the ancient mystery is the Changed Light's imprinting effect on the viewer or what I think of as the Shape Effects.

I believe that seeing the sight of the incredible fantastic muscular Man of Light and then seeing him stand up and turn around to his left and the fact that we humans are here on Earth looking up to meet his stare and go face to face with him is at the very root of the core of our joint human heritage.

5 RETURN VIEWER'S GUIDE #2

1. A massive forward rolling crater covered brightly illuminated moon coloured object rolls down from the area of space above and to the left of the Moon.

2. The object is traveling at a tremendous speed. Once the object enters the area within the distance of the Moon's orbit it will pass down towards the viewer's lower right and disappear below the Earth's horizon as it travels toward the area of space below the Sun in just a very few short minutes at the longest. Exactly how many minutes? I don't know exactly. The object is here within the distance of the Moon's orbit for a very short time as it crosses from the upper left to the lower right.

3. Although I have no idea how accurate this is I read that in ancient times using the naked eye the object was visible for a total of one hour from the time it was first sighted up till the point that it disappeared from view below down into the Earth's shadow and to the right of the Earth as it continues down towards the region below and beside the Sun.

4. I believe the object may be slightly taller the way it rolls. The object's surface has a giant mound area that I call the raised feature. As the object rolls forward this raised feature comes up from behind the object and travels over the top towards the viewer then rolls down towards the viewer and goes down under again on it's way around down behind and then it comes up from behind the object and then reemerges from over the top of the object once again as the object continues it's constant forward rolling action.

5. The object first covers the upper left corner of the Moon. The lower right area of the Moon is the last part of the Moon to be covered. The object crosses between the Earth and Moon at a very low angle. Because of this low crossing angle, the object has time to roll towards the viewer while continuing to eclipse the Moon.

6. During the time that the object is in front of the Moon before the raised area starts into view over the object's top or northern horizon I suppose the object's surface has been rolling towards you just the way you would expect a large sphere to do and look. Now the Moon is completely obscured. Up to this point the object's surface tilting effect towards the viewer has not happened yet. After the raised area has started to into view and then rolls forward beyond or past it's top or highest position the tilting effect has started or will start very shortly. The way the surface seems to stop rolling and instead looks like a flat area tilting over at you is the point when the valley is seen to the right. The object's surface tilting towards you effect instead of the object's surface rolling at you effect signals the beginning of the valley phase one of the Changed and Lasting Light event.

7. Even with a small telescope your viewing position is very up close to the massive boulder like object. My view started between five and maybe seven or eight seconds before the raised area emerged. At this point the Moon was not visible. The object does cross between the Earth and Moon at a low angle and the object does eclipse the Moon for a period of time. The eclipse does last. I am guessing that the Moon somehow could not have been completely covered for more than just a few seconds at this point. One of the factors that I base this on is the way my view started with me finding myself, looking basically looking at the middle central area of the rolling object.

8. The huge object gracefully rolls towards the viewer at the same time it was crossing in the telescope's viewfinder moving and going across from left to right and very gently slowly sinking. The object spends quite a few seconds rolling seemingly towards the viewer's position while between the Earth and Moon. This eclipse is not some sort of huge rolling object quickly zooming across your view of the Moon from the upper left to the lower right area in the view finder of the telescope as if it was just simply crossing from the upper left to the lower right of the viewer and then it's gone. Instead the object spends time rolling at you while it is in front of the Moon. The eclipse lasts.

9. Based on my guess that the full eclipse had probably started only a few seconds before my view started and based on my guess that only a few seconds passed from the time I first saw the object until the Moon's suddenly drastically Changed Light emerges and has become the sight of a fantastic immense valley, I am going to guess that the valley starts approximately eight or ten seconds after the object fully eclipses the Moon. It could very well be a few seconds more.

10. The object continues to basically roll at you. Eventually I did realize that the object was going to miss me once it had moved further to the right in the telescope's viewer eventually passing through the middle of the viewer and after it had dropped down lower. Looking through the telescope I had no sense of respective concerning many things including size, speed and distance. This is a very difficult thing for me to understand and explain. I do know from experience that looking through a telescope does effect how one perceives such things as the speed and the size and scale of such things as fast moving small planet size objects that are traveling by at an extremely close distance.

11. Before I bent over for my turn I suddenly noticed that it was as if a bright light had been switched on inside the telescope's view finder. The object's surface itself is so bright that as I was bending over for my next turn before I actually looked through the telescope I could clearly see the inside of the eyepiece and how it was constructed and how it looked all screwed together. The inside of the telescope's eyepiece was totally completely illuminated. After standing and waiting in the dark for my turn the sudden brightness when my view started was a shock for my eye. At the first few seconds at this early point I remembered thinking that my eye was not uncomfortable and I noticed that I could continue to look at the extreme brightness of the object's surface.

12. As the object's raised area comes over the top of the object towards the viewer the surface suddenly starts to seem as if it is tipping towards you instead of rolling at you. Look right and instantly notice that it is as if someone has magically clicked and dragged the edge of everything that you can see an incredible distance to the right way way to the right. Phase one of the Changed Light event has begun. It's the Moon but at this point the way the Moon looks has completely changed. Instead of the sight of the Moon you see a fantastic incredible valley! It is enclosed by mountains on all three sides. The mountains that go across the back of the valley are clearly visible in the extreme distance. Everything is oriented exactly the way you would expect to see some sort of very large

valley to look just the same way you would expect if you were looking at a real valley down here on the Earth.

13. As the object continues to roll forward the smooth rounded top of the raised area develops into a curving type of a vertical cliff. Gradually as the top of the raised area continues forward as the object continues to slowly sink the vertical area right before the top of the raised area, the curving cliff develops into an overhang. The area under the overhang is were the black spot of the shadow later suddenly develops This takes place in the far left of the overall scene.

14. Down between the mountains of the valley and down over the object's sloping upper right horizon area were you would expect to see some sort of valley floor you don't see anything at all that looks like a valley floor or ground at this early point. Instead an even level layer of an effect that looks like the top of a misty white cloud is seen. Everywhere in the valley is like the top of a flat white cloud. The surface or the flat area of what you would expect to see as the valley floor is indiscernible down over the object's far upper right slope or horizon and instead you see down into the area in between the mountains and the smooth level top of a misty white pure clean cloud layer is seen.

15. The light from the surface of the object seems as if it is spread wide open. The light from the raised area that eventually formed the overhang seems to somehow remain in the upper area of the scene and it's as if the raised area has stopped rolling towards you while at the same time the scene in the bottom of the viewfinder expands downward continuing towards you. The effect is like the light from the top of the scene is suddenly remaining in place while at the same time the light from the bottom of the scene continues forward and down. This effect seems to cause the object to suddenly remain in place and at the same time open wide open. A very noticeable valley effect is also now seen to the left in the viewer. The object's surface that precedes the raised area somehow now looks like it is a natural part of the valley as well. The way that the object looks is as if it is some sort of giant snail is extremely pronounced and somehow greatly exaggerated at this point before the black spot of the shadow has made it's sudden appearance. Even though the object's surface with it's fantastic craters on the left looks strikingly different than the sight of the immense valley to the right they seem to flow and blend together naturally and together the left and right become all the same place. At least that's how the effect appeared to me.

16. Suddenly and it startled me to see something happening. A large black spot appears clinging to the ceiling up under the newly formed overhang. At first I did not realize that the black spot was a shadow. I had never seen a shadow behaving this way before. Because of it's strange unfamiliar action, the way it spread outwards and thickened as it gathered in more ceiling area under the still growing overhang I just watched not knowing what it was.

17. As it turns out the Sun is basically directly behind and I think, slightly above the viewer at this point. The shadow is behaving normally and naturally but I simply had never seen a shadow cast in such an unusual way before. At this point the shadow just does not behave or look the way that you normally are used to seeing a shadow behave. Plus for me at the time with no foreknowledge concerning what I was about to see I was constantly being taken by surprise by every new sight and event that occurred.

18. The now very thick and strange looking black shadow very slowly at first starts to reach and creep a short distance down the cliff face right before it very suddenly drops or even pounces down to the valley floor below. Another unnatural looking action that startled me again! The shadow then immediately starts to race of to the right sweeping across the object's surface. Now suddenly the shadow looked and acted exactly the way that you would expect a shadow to appear and behave. At that point I remember feeling very relieved as I suddenly knew for the first time that I was looking at a shadow.

19. The wide and fast moving shadow looks like a massive black wave as it moves to the right and travels across the object's surface away from the viewer at a constantly increasing steeper angle. Early in the life of the wave of the shadow before the shadow's leading edge passes off the upper right area of the object's surface and into the light from the Moon there is time to look down to the ground. Looking left against the movement of the shadow's wave finding and focusing on one specific spot or area a fantastic effect is noticed.

This is the second time that I remember the effect of being drawn closer or seeing from closer to the ground up there. The first time being when the object tilts toward you and looks flat. At this point I saw the object from close up or similar to looking out of a remote camera up there. From that point forward continuing to see from up there now looking to the right into the Valley you can suddenly see from very close to the object's surface again.

As the shadow moves from the left to the right it reveals new surface details. When I looked down at this point I was suddenly seeing from a place that was right above the surface of the Object. At this point I was actually able to look out over the object's surface. Somehow the way the Changed Light works causes this effect to simply happen as long as you remain looking in an unbroken manner. This is the effect that I remember. I can't explain this effect and I did not at first realize this effect was happening when it was happening. I try to go into more detail on the page, The Shadow. This is the same effect that ultimately draws your eye's viewing position forward to the place were you suddenly find yourself looking down from above and behind at the sight of the back of the Awesome Good right before the Awesome Good turns to it's left and looks up right at you! A truly incredible moment. This is ultimately were the shadow leads your eye and this is ultimately definitely were you want to find yourself seeing from.

20. As the black wave of the shadow races towards the upper right horizon it also seems to start to thicken. As the shadow travels the tops of many crater rim pieces remain in the sun's light above the shadow. Also along the shadow's leading edge many small surface details seem to suddenly reveal themselves. Also shadows are seen crossing and traveling within individual craters. Definitely a feast for the eye.

21. The wave of the shadow does last but overall from the time the shadow actually drops and then crosses over the object's surface turning the brilliantly shinning object black until the shadow passes over and then off the object's upper right sloping horizon into the light of the Moon that is there, the valley only a very short period of time passes. Maybe only ten to fifteen seconds. I always struggle in the area of guessing the amount of time all of the different fast paced separate events take or last for.

22. Certain things are basically simple and easy to describe and many other things are very difficult to describe. Once the shadow pass over the object's far slope into the changed light from the Moon describing what things look like become suddenly more difficult compared to earlier when the shadow was still crossing the surface of the actual object.

23. As the wave of the shadow travels a series of tall individual shadows emerge beyond the shadows edge. These individual shadows travel

further and faster than the edge of the shadows wave. As the wave of the shadow passes over the object's upper right horizon these individual are clearly noticeable and are clearly defined.

24. At first the individual shadows are short. I know that scale is very hard to judge and as a result the way these individual shadows looked to me or rather what they reminded me of changed as I glanced to the left away from the series of individual shadow and then to the right back towards the individual shadows.

25. Once the wave of the shadow has passed over the surface of the object and has reached the valley area that is the changed light from the Moon the separate individual shadows that precede the wave of the shadow reach further faster into the changed light from the surface of the Moon and it's as if these individual shadows are all now traveling on a flat smooth surface.

26. At first while these individual shadows were relatively short they looked very much like a fence stretched across a piece of property. Somehow the changed light from the Moon's surface appears or arrives simultaneously when the points of these individual shadows arrive. As these individual shadows lengthen they travel over a flat smooth area that is just somehow suddenly there for the shadows to travel over. This flat smooth ground or area forms or builds out away from the viewer farther and farther transforming the wide flat white cloudy misty area that is surrounded by the steep faced mountains of the valley as it travels. A very incredible sight.

27. When I saw the individual shadows for the first time I did not actually realize that they were lengthening or traveling away from me. I was looking all around but mainly I was looking to the right and to the left. If I had remained focused on the individual shadows I would have seen them growing longer as they traveled away from me. Instead I looked back to the left away from the individual shadows and then I looked back to the right again and saw these individual shadows at a later longer taller stage in their growth or evolution. Now instead of looking like a series of fence posts across a yard or property they looks exactly like telephone poles going out across the flat countryside. You see down to this scene from above.

28. In between the time that these individual shadows are short having first emerged right after the big main wave of the shadow has finished

passing over and then off the object down over the far upper right slope into or onto the area of the flat white cloud effect and up to and perhaps just slightly past the time that these individual shadows look like a line of telephone poles there is some time to look at the far upper right slope or horizon of the object. I think of and call this area the far slope.

29. It turns out that looking away from the individual shadows and the main shadows goes against the big main point of it all but there is a few seconds a very few seconds available to the viewer to look over and at the fantastic scene that is there to be seen down over and actually above the far slope. A very important point to remember is that you want to find the point of the biggest tallest widest shadow so that you can let your eye be drawn deeper into the changed light from the Moon. At this point in time that is the big main most important goal. Finding this biggest tallest widest obelisk shaped shadow's point and letting it draw your eye along with it will allow you to arrive at the place were you see down from just above and from just behind very close to the ground at the exact instant in time that the back of the Awesome Good is first seen suddenly formed having just been created sculpted by the point of the shadow that was drawing your eye's viewing position forward. Arriving at this point is obviously every viewer's main most important goal! If the point of the big main shadow strikes and starts to draw the squiggly line and you are caught of guard looking to the left, the wrong way... let's just say that you don't want that to happen to you...!

30. Everyone should glance at the far slope. It is a sight that is not to be missed but it is not the biggest most important sight that is to be seen, plain and simple although this sight is important for many reasons. After the wave of the shadow passes over the object and the sudden thickness of the shadow is seen over the object's upper right horizon there is time to look at the fantastic brilliantly illuminated shapes that are suddenly seen perfectly arranged on top of and along the line of the top of the shadow. An incredible fantastic scene! I use to think that these incredible shapes were just the tops of some of the standing crater rim pieces and this may be true.

Now I am wondering if these brilliantly illuminated shapes could be a combination of the tops of crater rim pieces as well as the light from stars in deep space in the background that is seen over the objects surface. In particular there are two attention grabbing triangle shapes that are the two most prominent and largest shapes that are seen on the top of the thickness of the very black shadow's line. Are these incredible shapes a combination of the tops of crater rim pieces as well as background stars? Recently I have heard about the way that planets come into some sort of alignment or grouping. Could some of these incredible shapes that are seen above the shadow down over the object's far slope actually be planets? What about the new planet that was recently discovered? How does that body and or other similar types of perhaps undiscovered bodies figure into this situation? What does the big planetary alignment picture really look like? Time will tell. All of those big important experts out there are really going to love this one! I can't wait to hear what they will have to say.

31. Anyways the point is look at the shapes above the shadow over the far slope, but don't stay looking at the far slope for to long. If you do stay looking at this amazing scene for more than just a very few seconds it could and will cost you big time. It's like watching the big game. You can be watching and intensely looking and at the same time you can still miss the big play.

32. The most important point is out of all of the individual tall narrow obelisk shaped shadows there is a single most important tallest biggest widest most important obelisk shaped shadow. Every viewer needs to locate this shadow at the earliest point possible and every viewer has to focus on the very point of this all important shadow so that they will arrive at that very special place were you suddenly see down from above and behind so that they can see the big sight of the Awesome Good. Then if they you are suddenly looking down at the back of the Awesome Good then split seconds later it will be your turn and you will suddenly find yourself going face to face with the actual Big Thing the intensely mind bending impossible sight of the Awesome Good, the ancient muscular Man of Light.

33. Also if you are looking to the left in the scene and you are looking at the actual object itself be aware of this, the object will have traveled further down as well it will be further to the lower right. The light from the Moon and perhaps even some of the light from the object remains above the forward rolling object. The raised surface feature that earlier formed the smooth cliff and then the overhang has rolled farther forward and down. The sinking object is literally falling away from the face of the Moon at this point although the object's light remains above the actual horizon of the object, mixed and blended with the arriving light from the Moon.

This is the point when the object reaches a certain lower point were it starts to look more like the object that you saw before the view of the valley presented itself. This very unusual visual effect is hard to describe however you will know exactly what I am describing as soon as you see this effect were you suddenly see the wheel turning again. Suddenly you will be able to see the normal round or spherical shape of the ball of the object again.

34. Every viewer needs to know these next important points. These points that follow are important because every viewer needs to know when the last few seconds are ticking down.

35. These next points signal that time is about to run out. If you see a great expanse of the object's normal spherical curving surface area suddenly visible over and past and beyond the raised surface area that cast the all important shadow in the first place then time has very nearly actually run out. If you see this indicator you are looking to the left in the scene the wrong way at the wrong thing. You are looking at the hard ground of the object. You are looking to the left.

36. Instead of looking left you really really need to look way way back to the right. Look to the right deep into the distance of the changed light and find the point of the big main central shadow because it is nearly time for the first of the ultimate sights. There are very literally only seconds left and you need to know that there is no more time for sight seeing and instead it's time to focus and brace yourself for what is about to happen before your eyes in and then very suddenly literally out of the changed light from the Moon! You don't want to be this close and miss the first instant because you were looking to the left instead of to the right! Look to the right ... Now!!!

37. THIS IS CRITICALLY IMPORTANT! Again if you are looking left at the forward rolling crossing sinking object itself and you find yourself suddenly seeing over and past the object's raised surface area the overhang and you are looking at normal surface area as described directly above you have only a very few seconds left to look back to the right so that you will be looking the right way in time to see the sight of the Awesome Good suddenly formed! Look right! Perhaps glance at the far slope for a split second on your way back to the right as a way to ensure that you see the two brilliant triangle shapes in the foreground of this area. It may turn out that these two triangle shapes the largest most brilliant attention grabbing sights seen above the far slope may not make

an appearance or present them selves until a later stage in the evolution of this event. However at this point you really are taking a big chance if you linger. I would recommend against looking at the far slope for more than a split second at this point. I can't be exactly sure just how many fractions of a second are available for sight seeing in the region of the far slope at this point. I saw this many years ago and trying to recall the timing of events down to the last split seconds at this critical point is very difficult.

38. Once you can see over and past the top of the overhang the raised area, you do have time to glance at the far slope as your eye travels to the right but that's all the time that's available for viewing the shapes above the shadow down the far slope at this point. I can guarantee these preceding points. There may well be a few more seconds for sight seeing at this point I'm not sure. I can only strongly suggest to you that the time for casually looking around is over. Hurry hurry hurry and get busy looking to the right and find and focus on the the big main obelisk shaped shadow's point! Notice that from nearly directly below your eye's viewing position you will see the wide line of the big pointy and straight main obelisk shaped black shadow now very rapidly racing into the distance of the changed light from the Moon.

39. A main most important group of three shadows have emerged and together they travel farther faster than the other individual main shadows. At this point I don't even remember being able to even see the other individual shadows. It really was as if the view of these three main shadows was somehow magnified and the effect was they were the only shadows that I could see. It's a natural funneling focusing effect that happens as long as you are actually looking. It may be that if you are not looking into the right area within the light, you may not find yourself drawn forward up, to were you see down from.

A very difficult effect to describe in spite of the fact that this effect is so vividly and easily remembered. Very soon everyone who also sees, will know exactly what I mean. If you are looking to the right and you are using a telescope you should be able to easily find the big main shadow because as I describe at this point everything just seems to start to happen naturally. I think that if you are looking to the right it would actually be hard to miss the sights I describe. You will not find yourself doing any sort of searching for the big main shadow at this point instead if you are looking right, the light will allow you to see because of the focusing funneling effect that I describe.

40. Now at this point the position that your eye is seeing from is lowering down closer to the ground along with seemingly following along behind the three main shadows as if being drawn forward by or even riding forward in the light. Also everything that is out wider that is a part of your peripheral vision seems to naturally follow the flow forward. Now you will have noticed that during the last few seconds, the flat ground is closer below you as your eye's viewing position seems to race seemingly faster, farther forward as it keeps gradually getting closer to the ground.

41. I really don't know if the point of the big main central shadow is actually accelerating at this point or not. It is definitely moving away very rapidly but is it also accelerating? I just don't know. I think of it as accelerating but I don't actually know if the point of the big main black shadow does accelerate or if it is just simply moving very fast at this late stage.

42. By now everyone needs to have found the big main shadow so that they will be able to see the first instant in the Moons changed light. Do not look away! You would think that not looking away would be simple. Brace yourself for the very complicated difficult sights that you will be seeing very shortly! At this point there is nothing more important that focusing on the point of the big main shadow so that you are ready to see down from above and behind at the sight of the muscular back of the Man of Light. This really is the place were you want and even need to be seeing from! There are now fewer split seconds left... Time now runs out!

43. This is it! Suddenly the central taller longer wider shadow's point goes up and down and up and down then instantly the shadow's point cuts zig zagging to the left and to the right and it starts to draw a squiggly line that is the shaping and the forming and the sculpting and the creating that produces the sight of the muscular back of the Awesome Good!!!

44. This really is it! The first great sight! Your eye has arrived at the place were you see down from! Look down and see the clear awesome sight of the very muscular back of the Ancient Man of Light. Fantastic!

45. This moment only lasts a very few short seconds before it passes and it's over. An incredibly peaceful and tranquil moment of thought, incredible!

46. Our ancient ancestors knew this sight and I can only suppose that this is why we all also know this sight and this pose as the thinker. Sitting or crouching facing away towards the distance of the Valley you see his back and his right side. This is obviously were this ancient depiction comes from. Enjoy this very very brief sight because this is all about to end in the most stunning shocking manner!

Shortly after the next few seconds you will be suddenly even physically forced to realized what you are suddenly seeing. This is the point that is most difficult to actually realize because of the impossibly shocking nature of the next sight. At the same time knowing ahead of time that you are not suppose to feel extreme shock and fear may very well help you at this point when you really do see the heart of the changed light for the first time!

This is only my opinion but if a person is able to remember that somehow it's all about the opposite of shock and fear then perhaps they may be able to experience less shock and fear during their own view of this event when they are actually really seeing and also importantly later after their view of this event is over when they are remembering this sight the special way that the shape effects work throughout all of this.
It may turn out that the more you are able to feel and think in terms of overwhelming joy and love during those very personal face to face seconds of the Big Noticing the more likely it may be that those are the feelings that you will instantly feel whenever you are actually really thinking about this experience again and this face to face sight later when very suddenly you see and feel that the shape effects are at work.

47. I do and at the same time I don't actually know this because instead for me it's the extreme opposite situation. I had no idea that it's all about extreme joy and love. Instead I had the extreme opposite reaction when I saw and went face to face with the Man of Light and as a result unfortunately for me in the past when I remembered the special way that the shape effects allow you to remember I would suddenly find myself very involuntarily feeling the exact opposite way that you are suppose to feel. A very difficult effect to describe but I believe this effect is real.

This is very obvious to me, I am hoping that there may be a chance that mentioning this very difficult and personal aspect of this experience may help others to know and experience the fantastic face to face seconds the way I truly believe they were intended to be by nature. For me being this open and honest is difficult. As a part of trying to do and say what is right I clearly understand that I have to let the things I know lead me even when some of those things lead directly into personal areas that are difficult to openly talk about. If it turns out that including the above points and comments actually helps people to know this experience the way I think it may be intended to be experienced then I am glad.

48. The peaceful tranquil few seconds that go by while you look down from above and behind at the back of the Awesome Good very suddenly end! Basically you have only had enough time while looking down from above and behind to realize what it is you are looking at down there before suddenly everything that you are seeing drastically changes. I am able to read about many other big moments that follow this particular point in time but except for a few the rest of the big moments or big sights all happen later during phase three of the changed light event after the changed light arrives at or apparently very near the Earth. The face to face moment of the Big Noticing is easily the biggest most fantastic and most difficult and wonderful sight that I have ever seen.

49. My viewers guide will have hopefully helped you to arrive at this incredible fantastic place in time and space. Now it's very much up to you to keep looking so that you can go as far past this point as possible. I was very fortunate to arrive here so that I could experience the actual face to face moment. If you are finding yourself looking at the back of the Awesome Good then you are also going to suddenly be noticed! A very fantastic moment when you see the most spectacular sight. I hope that the big main point that I make from this special moment onwards will be of some help. Simply put keep looking! Don't look away! Focus on forcing yourself to return the intensity of the stare! Remember the part about joy and love!

50. Suddenly in only a split second the Awesome Good turns to his left, spins to the left and gets up all in one fast motion and is suddenly oriented directly at you and the ultimate face to face staring contest is on!!!

51. A number of heart stopping seconds go by as this close face to face seconds last! During this time the next one way arriving wave of light moves up in your direction from the Moon and this causes the face to face to drift up on an angle closer to where you are looking down from. The distance closes as the Awesome Good rises and nears you! Definitely incredible wonderful and fantastic come nowhere close to describing it! Suddenly these close up face to face seconds end.

52. Now the very bizarre transition period after the valley phase and before the mountain phase occurs. The outer edges of the light from the valley pulse inwards a number of times. The Man of Light himself wiggles around and orients himself differently and he is evolving at this point into his second evolution. A very bizarre sight that I can describe in detail. Keep staring back and don't let the continuing face to face staring contest that you are locked into at this point end!

53. The place were you see from has raced back during the light's inward pulsing then almost instantly the intense pressure and force of the lasting light's focus arrives and is felt right after the light's inward pulsing has stopped. The difficulty for the viewer constantly keeps going straight up!

54. Everything has suddenly sped up and then incredibly everything that you see will be rapidly getting closer to as once again incredibly the intensity and the pressure of the light's focusing stare starts to build higher and literally rapidly rise right up through space as the actual Big Thing the Man of Light seems to be noticing and staring at you even more intensely as he rises.

Stay looking although now changed the incredible face to face sight of the Awesome Good is still happening and is rapidly nearing and closing in on your exact position from down inside the incredible growing bending curving mountain of changed Moon light as every viewer will very soon suddenly see and realize.

Stay looking! Don't look away! The longer you can see for the better you will feel about it later. My sincerest best wishes to you and your family from me and my family! Good Luck!

6 RETURN VIEWER'S GUIDE #3

1. When I try to describe the object's motion and direction it is always from my own viewing position and perspective. The place were I stood in Mississauga Ontario Canada. The forward rolling crater covered object arrives from the area of space above and to the left of the Moon. The first part of the Moon that is covered by the object is the Moon's upper left corner. The object appears to roll down from that area of space. Overall the object is moving at a tremendous speed and is between the Earth and Moon for only a short time. Because the object's crossing angle is very low, the object is between the Earth and Moon rolling towards the Earth for a fairly long period of time relative to it's very high speed.

2. Although huge and massive the object is probably smaller than the Moon. I understand that something that is smaller than the Moon can still in fact be very large and massive. The object is easily in fact a very very big place. The object crosses the line of sight at a position between the Earth and Moon well out in front of the Moon. This crossing point and angle allows the object to completely obscure what is probably the larger Moon for a many seconds as the object continues to roll towards you. A normal in focus view of the Moon is much less clear than the brilliantly shinning incredibly clear extremely in focus view of the object and the object's surface details.

3. The object's crossing angle is so low that it certainly does appear to the viewer that the object is rolling basically in there direction. The object does cross at a point that must be a great distance well out from the Earth as well as the Moon. Through a telescope the object appears to be very /

close and bearing down on the viewer as it seems as though it surly must just be somewhere close and directly overhead. The telescope places your eye's viewing perspective at a point that is well out from the Earth. Your sense of being at a safe distance back from the forward rolling object is completely gone as the extremely bright massive rolling place fills the telescope's eye piece.

4. The object is slowly sinking as it moves from the left to the right in the viewer. The object's left to right crossing rate of speed is higher compared to it's much slower downward sinking rate of speed. An incredible hair raising sight that is guaranteed to get all of your adrenaline pumping! Even though the object does miss the obvious startling feeling that you get from viewing through a telescope is that basically it's headed for you. Looking through a telescope if you did not know ahead of time that the object does in fact miss the Earth you would think that surly it was definitely going to hit. For me at the time not only did I think that it was going to hit the Earth at first I figured it was actually going to hit my neighborhood!

 The first point that I had any sense at all that the big rolling place was going to miss was when it finally started to fall away from the Moon down to my lower right once it had passed the center of the view through the telescope. At that time however as a kid I only knew that it had actually missed the Earth completely once days had pasted and nothing seemed to have happened. I know that this has got to sound completely ridiculous but I was a kid looking through a telescope that night for the first time and I had basically less than zero overall understanding. The view through the telescope caused any and all sense of perspective to be completely thrown off. I had no idea or sense that the object was actually far away and crossing at a point well out from the earth. I am easily able to understand that this basic point is an important point to know and may actually help future viewers during these very stressful moments that are approaching us right now.

5 All future viewers will know that their telescopes are still pointed at the Moon. That is something that I did not know when I bent down and saw the object for the first time. Instantly I thought the telescope had been moved. While I was seeing I had no idea that the Moon was about to emerge from over the object's right area and upper right horizon. I try to take time as I write to explain how various factors like the one I just mentioned helped to cause my and contributed to my lack of perspective and my very very limited understanding concerning what I was seeing

when I was seeing it. I what to try to ensure that future viewer's are in a better position than I was concerning understanding what they are seeing when they are seeing it. I can only do so much but I am going to keep trying my best to help future viewers. Looking through a telescope throws off your sense of perspective concerning the size and distance and the speed of the object and the Moon's light. That is unavoidable. What is avoidable is future viewer's having a complete lack of perspective and understanding when they are watching the return events unfold. A lot happens in a very short time. The light that you know from the Moon becomes changed and unrecognizable and it will confuse and fool you. I do understand that I am able to help people understand some very basic things now that I understand these basic things better.

6. Except for the left side and perhaps the right side of the forward rolling object I did not see or rather I could not see the actual right side of the object as I try to explain. The surface of the object is completely covered with even hundreds of the most fantastic and unique craters. Individual vertical standing tall narrow crater rim pieces very neatly circle all of the various size craters. The standing crater rim pieces have a very uniform look to them. Their thickness may reflect the thickness of the object's crust.

7. The craters themselves are not some sort of deep hole the way you would normally picture a crater like for instance a typical lunar crater. Instead all the craters that I noticed and took a look down to were no deeper than the thickness of the crater rim pieces. It may be possible that the area inside the circle of the craters may be basically flush and level with the object's uncratered surface area. I am unable to be more precise. My guess is that the craters are as deep as the crust's thickness, the thickness of the standing crater rim pieces. I know that it is one or the other. I remember this detail and I see this detail being a certain way. I do realize that I am pushing my memory to it's limit in this case and I really don't know which situation is the case.

The sight of the crater's are real and not a shape or form made out of changed light. As a result I have to try to simply remember the details concerning the depth of the craters as I describe basically without the benefit of the way the shape effects lets you remember details. Aside from this area concerning the depth of the craters were I have it narrowed down, the rest of the descriptions I give concerning the craters seen on the object's surface are completely accurate guaranteed. If you've ever seen a standing stone circle down here on Earth and we all have then basically you've also seen a crater on the returning object and that's also guaranteed!

8. Once the sight of the Moon is obscured by the object all you see is the massive crater covered boulder very gracefully rolling at you. Suddenly a large raised area comes into view from over the top of the forward rolling object. Future viewers will have already seen the raised area coming over the top of the object towards then as they will be looking at the object before it crosses between the Earth and Moon and then blocks out or eclipses the Moon.

Does a shadow develop with each of these earlier rolls as the object approaches it's position between the Earth and Moon? I don't know because my view of the returning object started after the raised area had gone down and around and was getting ready to once again come up from over the top again once the object was completely obscuring the Moon. It could be possible that from whatever point onward once the object has come down low enough a shadow may be cast from under the overhang that the raised area becomes with every forward rolling revolution that the object makes before the object arrives in place in front of the Moon or maybe not I don't know we shall see.

9. My view started seconds before the object seemed to magically settle into place in front of the Moon. Now I understand that this settling in starts when the light blending between the object and the Moon starts to happen. At first I saw the object rolling down without any sort of place attached to the right. A very fast paced series of events take place. Once the object is in front of the Moon each part of what you are seeing is an area or time segment or a sight that lasts for only a few seconds. Some things take longer to happen that others but the things that happen and the separate sights only last for short seconds. I try to estimate and remember how long the various sights lasted. It has to be remembered that I can't be completely accurate down to the last second. That sort of accuracy guessing time and counting seconds from so long ago is something that is impossible to do with complete accuracy but it's important to try because understanding the timing of the sequence of events is important.

I also understand that the shocking nature of the experience after and while the Return events are happening makes judging time down to exact seconds very difficult. I have to say that I think that I am basically fairly accurate overall. Being accurate overall does not mean 100% accuracy.

There are sights that last for split seconds they last for split seconds. Things and sights that last for a much longer time are much harder to accurately estimate.

10. Suddenly the large raised area comes into view from over the top of the forward rolling slowly sinking crossing object. The object's entire surface area still seems to be rolling toward you just the way that you would expect. Then suddenly the area before the high raised area seems to suddenly be flattish looking and tilting toward you as you watch instead of the normal rolling toward you that you had been watching up to this point. The sudden tilting is the point were the Moon is suddenly seen to the right and to the upper right past the edge of the object's rolling sinking horizon. The light from the object's right and especially the edge of the object's upper right horizon is suddenly blended together with the now changed light from the Moon. The incredible sight of an immense valley is seen to the right. It's the Moon but it does not look at all like the Moon anymore. Instead a fantastically clear view of a very real looking valley is seen.

11. Steep mountains are seen on both sides with a wall of mountains going across in the distance. A white cloudy misty look is seen were you would expect to see the valley floor. This white misty cloudy look is seen valley wide and covers the entire valley floor area from the edge of the object's upper right downward sloping horizon right up to the edges of the mountain's smallish foothill type of look that is seen at the base of the steep face mountains.

12. As the object continues to roll forward and slowly sink and move from the upper left to the lower right in the telescope's viewer the very smooth and curving raised area begins to gradually form a smooth curving vertical rounded at the top rock cliff. Then an overhang gradually forms as the object's raised surface feature continues to travel over forward and then down towards the viewer.

13. The Sun and the Earth and the Object and the Moon are all lining up. As the big raised area continues forward it continues to form and develop into more of an overhang. Very suddenly a very noticeable very black shadow appears up under the overhang. A very big and very noticeable black spot. I think of it and refer to it as the black spot of the shadow.

14. The black spot that I didn't realize was a shadow when it first appeared instantly starts to progressively gather in more of the ceiling

area under the overhang to match the object's forward moving forward rolling sinking motion. During this time the shadow thickens and or gets deeper I suppose depending on how you look at it or depending on how it happens to look to you. A very bizarre unnaturally behaving shadow. I suppose the shadow seemed unnatural because I had simply never seen a shadow formed in this way before.

15. The black spot of the shadow's sudden appearance instantly caught my attention when it appeared in the left area in the overall scene. The black spot of the shadow spreads and thickens. Suddenly the shadow seems as if it leaps downward to the valley floor below. At this point even though it's as if the shadow has dropped down to the left end of the valley the shadow is actually still being cased on the actual surface of the object itself by the standing stone craters. The shadow now an awesome wave still has to cross over the object's surface before it actually reaches the object's upper right horizon the area I think of and call the far slope. This takes some time. The shadow does travel fast but overall the shadow does has a considerable distance to travel.

16. Once the shadow reaches over and off the object's far slope into and onto the valley floor area of flat white clouds somehow the the light from the Moon where the flat white clouds are located transforms and instead now looks like an area of smooth flat hard rock as it meets or is touched by the leading parts of the shadow's edge. A very natural transition between the actual hard surface of the object and the light from the Moon is seen. I was looking left and right and I did not follow and look at the actual edge of the shadow's wave in an uninterrupted manner. Some things are unclear to me because of the way I was looking back and forth to the left and right

17. Once the wave of the shadow has passed over the far slope into the valley area that is surrounded by the mountains there is a fantastic scene above the thickness of the line of the shadow. Many many shapes are brilliantly illuminated above the thickness of the black shadow. I think that the tops of crater rim pieces that remained above the depth of the shadow reflecting the sun's light account for the majority of the fantastic brilliant shapes that are seen down the far slope. In a very bizarre way these shapes look. like the most incredible cityscape with everything situated and orientated facing and looking out into the distance of the valley just like you are.

Another strange effect is that even though obviously the object itself is still moving and rolling and yet at the point when you are seeing the shapes above the line of the shadow the object itself is already actually gone down lower and to the right. The brilliant shapes that are seen down the far slope remain unmoving and lasting in place. As a part of the light blending and the effects of the changed light somehow the light reflected from the tops of crater rim pieces that are gone is still there. Again obviously the craters are continuing down and are literally down below and gone but the light that reflected off the tops of the rim pieces is is still there and now a natural part of the scene.

18. Also it could be possible that the light from back ground stars and perhaps even planets that may be in some sort of grouping could also contribute to the what the viewer sees. Maybe once the surface of the object has dropped down to a lower point light from more stars may suddenly emerge and become part of the fantastic array of brilliant shapes seen down over the far slope. In particular there are two fantastic triangle shapes that are seen within this area. I am uncertain in regards to the specific timing of their emergence.

Were these two triangle shapes visible or there and developing or evolving at the same point that all of the rest of these brilliant shapes were first seen above the shadow?

Did these two triangle shapes appear at a later point after the first group of shapes are visible above the shadow?

I don't know because I was looking left and right and even up but especially down to the ground up there. My feeling is that they appeared later after the first group of shapes above the shadow. I also think that because of the way they really were extra stunning and larger than the other shapes and because of their possible later appearances they may well be a pair of stars instead of reflected light from the tops of crater rim pieces. I don't know this, instead this is a feeling, a guess. Plus there's that Orion business that I hear about. Later we will see how all of this actually works.

19. Overall once the shadow has passed over and off the object into the light of the Moon the valley there is very little time left to do anything other than focusing on finding the big main shadows point. Finding and following the big main shadows point is the most important task for the viewer at this point. The seconds are very quickly counting down and the big point in time when you are going to be looking down from above and

behind at the very end of the black wide long road that you have been hopefully traveling on is approaching very fast. Time is running out! Look to the right and find the main group of three shadows that are traveling together. The middle shadow is wider and longer or taller than the two outer shadows that trail behind the main central shadow. Find and focus on the very point of the big main shadow! It does travel on a slight angle even though it is actually straight.

20. The object is crossing from the left to the right out in front of you between the Earth and Moon. The Sun the actual real Sun is positioned behind the viewer behind the Earth in line with the Moon and the object and the Earth. At first the shadow travels from left to right and at the same time the shadow is traveling on an angle away from you that is steadily increasing as the shadows lengthen. Initially there is a lot of left to right motion. Later as the lines of the shadow grow very long the angle they are seen at has increased greatly and now the shadow is lengthening and traveling almost all the way up to ninety degrees or straight away from you. As the object continues to cross the shadow goes from having a lot of left to right motion all the way to nearly traveling straight away from you without any motion from the left to the right at all. The lines of the three main shadows become straight as they pass over the flat whiteness of the valley turning the the flat white cloudy mistiness into flat smooth hard ground.

21. Does the shadow reach all the way to ninety degrees before it starts to go up and down and up and down? I don't know. It might, it might not. I don't really think that it does but I can't be 100% certain. Events start to happen very fast at this point. This is the point were the viewer, at least this viewer, started to lose track of the details of exactly what's happening. My viewing position was racing forward as if zooming like a zoom lens up into the distance. I was less aware of the surrounding details all the time as the point of the big main shadow dragged the place I was seeing from further forward into the distance of the Moon's changed light. In general I hope that my descriptions can help future viewer's find their way to this point. The exact angle of the shadow is really not important at this point. The point is to find and stare at the very end of the tallest widest shadow that is in the right side of the entire scene that you are able to see. That's what's important!

22. I spend time mentioning smaller details and points because I understand that mentioning these points may turn out to be useful for future viewers. Not because the small details are important or critical to

45

being able to find the big main shadow. If you are looking the correct way to the right at the right time once the shadow has traveled over and off the far slope you will see the big main shadow! It will just simply be there right in front of you for you to follow.

23. If it turns out that the line of the big main shadow goes past ninety degrees before it strikes, and it might, who cares? This is a general viewer's guide. I will be out or not 100% completely accurate some of the time. Being completely 100% accurate concerning every detail all the time is not possible. In general following my viewing tips and suggestions will get you to the ultimate ancient place in the Moon's changed light when it's time to look down from above and behind and it's nearly time for the Awesome Good itself to turn around and get up and notice you. This is were you want to be! Me being out a few degrees here, and a few degrees there will make zero difference. Follow the points that I suggest and you will get to see face to face with the Man of Light.

24. I mention the idea about the timing of events. I mention about seeing over and past the object's raised area and how it is important to know that on the left if you can actually see over the object's raised area and you can see the normal round spherical shape of the object's surface again then there are only seconds left before the point of the big main shadow starts to sculpt the form of the Awesome Good in the right area or the upper right area of the valley. This is actually important to know but if you are already following the wave of the shadow into the distance to the right and you are not looking back to the left and you don't plan to look back to the left then suddenly seeing over the top of the object's raised area is not important to you. You will be already looking to the right and by this point you will have easily already found the big main shadow and so whatever is going on way back to the left will not matter to you. You will already be looking at the single most important sight that there is to see at this point.

25. To the left the shadow appears under the newly formed overhang. Then suddenly the shadow drops to the surface of the object below. Then instantly the shadow races from the left to the right over and off the object's surface towards and into the upper right area. Tall individual shadows separate themselves out ahead of the big black wave of the shadow. Find and focus on the point of the big main tallest shadow and don't look away as it grows longer and longer and then suddenly your eye will arrive at the place were you see down from and then you will

see the muscular back of the Man of Light. Then the Awesome Good turns around and you will be actually going face to face with a solid looking shape that was made out of changed and lasting moon light by the point of the big main shadow. It's basically very simple follow the shadow from the left to the right and you will see the ancient repeating sights that have always happened! Then beyond this point it's all about not looking away! It's important to remain looking in an uninterpreted manner. Stay looking and you will see more and as a result you will know more! Everything speeds up and the rising changed light's intense pressure arrives and builds and then everything goes from there! Be happy be glad and try to remember the part about joy and love. In the end joy and love matter! Once the face to face situation is in place it keeps happening! The inward pulsing starts and ends and then everything that is light just simply rises as you continue to go face to face with the Man of Light!

Don't look away! Keep forcing yourself to receive the light's indescribable pressure and very difficult intensity! There are a lot of details that are worth mentioning but overall it's about looking at the right place at the right time. It's a natural!

7 RETURN VIEWER'S GUIDE #4

1. We were using a small backyard type of telescope. The telescope was a new Tasco 3" achromat a popular model at that time.

2. We were using whatever eye piece gave us the overall clearest view of the Moon. It is my opinion that this overall clear view of the Moon contributed to the crystal clear very in focus view of the object and it's surface, and the extremely clear sight of the Moon's changed light.

3. Basically I think that you don't want to be in really close to the Moon. I believe an overview type of viewing perspective will prove to be best overall as opposed to a very tight in close view of the Moon. Once again this is only my opinion. If a person had the luxury of more than one telescope then perhaps both the in close and an overview type of viewing perspective could be used to great advantage. I think that this would only be the case if the viewer understood the fast paced timing of the sequence of events.

4. The overview type of viewing perspective is the type of view I had. This allowed me to see the overall scene. Once the object had settled into place in front of the Moon I was able to see space to the left of the object and I was also able to see over the top of the object.

5. With the Moon positioned slightly to the right in the viewer everything seemed to be framed perfectly. The object on the left and the view of the Moon, the background for the valley on the right. If the Moon had been perfectly centered in the viewer then there could be a chance that the object's left horizon the left side of the object along with some of the

object's surface may not have been visible to me.

6. The higher power small backyard telescope that provides a much closer look at the Moon may not give a viewer a chance to see the overall scene quite as well as the lower power telescope. On the other hand a closer tighter view of the Moon itself would allow for a very in close view of the spot were the focal point of the illusion forms.

7. For me personally I understand that the overview type of view provides an extremely in close view of this incredible event and the incredible sights that form there. At that very special point in time your viewing perspective seems to have raced forward and you seem to be looking down at the sight of the Awesome Good from an extremely close viewing position. Defiantly for me at least this is close enough! Seeing from even closer is something I think of as being the realm of only the very brave!

8. The idea of switching from a lower power telescope to a higher power telescope once the time is right is something I have thought of but I seriously doubt that I will do.

9. The series of events moves at such a rapid pace that if you are not sure the timing you could get it wrong and you could actually miss the ultimate sight at that ultimate instant in time. Defiantly viewers don't want this to happen to them.

10. My main advice is once you are watching the shadow's awesome wave and you have located the big main obelisk shaped shadow you should remain focused on the idea of concentrating on the point of the big main shadow. The idea of stopping to look through a different telescope, could turn out to be your ultimate life's regret. I can only recommend against it.

Once you are actually seeing the object's awesome surface with the shadow's wave moving across it keep looking. You are almost there and very shortly you will be seeing the awesome visual spectacle that follows. Don't look away! Locating the big main shadow in the middle of the three main shadows is a natural once you are looking to the right, in the right region at the right time.

11. In my opinion, in the end or is it actually the beginning when the Moon's light very suddenly focuses on you remaining looking will be

one of the most difficult things that you will ever attempt to do. I do think that if a viewer is prepared and actually knows ahead of time that they will need to brace for the big moment, and that there is a big moment, then remaining looking is something that they will be able to do, I hope. I understand that switching telescopes should only be done early in the life of the shadow and not later.

When future viewer's are watching the event unfold I hope they will have some sort of sense of were they are within the sequence of events based on my descriptions. Before the shadow it's safe to switch telescopes. Early in the life of the shadow it's also safe to switch telescopes. The shadow has a very short life. For most people there may only be one chance to get this right. Late in the life of the shadow switching telescopes for some sort of closer look could cause you to miss the sight of the big moment. Stay looking and don't let that happen to you.

12. During the last return I had no clue that I was about to see the ultimate sight. I was a young boy with zero experience, no lead up, no preparation, and completely unsuspecting. I looked down through my friend's new telescope again after I had looked at least once before as I know I had seen the Moon by this time at least once. As if someone had changed the channel without me knowing, the massive object was just simply suddenly there. At that point I had no clue that the telescope was actually still pointed at the Moon as I instantly presumed it had been moved and pointed at something else, as if my friends Dad had pointed the telescope at the next big thing we were going to look at. Right from the start any sort of perspective or understanding was totally completely thrown off. I was already instantly struggling badly with my first peak at the very intimidating massive crater covered boulder the object, that seemed to be basically rolling right at me!

13. That was just the start. I have no idea why I was not already running even at that very early point in time. Like me, surely most or all telescope aided viewers will defiantly have that very uncomfortable caught in the middle of the road feeling! I think that because I knew so little and because I did not know that such a sight was actually unusual, somehow I stayed looking from this initial early point right through the series of visual events that takes place all the way through the big instant in time. I also managed to stay looking for a time past the big instant but very unfortunately my view would be soon rapidly nearing an end as I describe. Completely shocked and totally confused and overloaded, and

badly struggling comes no were close to describing it.

14. Ultimate seconds that everyone should know! It turns out that people can know these seconds and I understand that they will know these seconds. The impossible, the indescribable sight will be there happening again, soon! People need to understand that they can see it to. If they see, then they to will know.

15. In the end it's all about seeing that's the point! The only way that's going to happen is to look and to stay looking. I understand that this can only sound impossible but the simple fact is that unbelievably actually remaining looking is not easy. I try to describe the shocking pressure and strange intensity that is really felt by the viewer. The full force of this pressure and intensity arrives with the second phase of the lasting light event, the rising bending incredible lasting light mountain that happens right after the the inward pulsing of the transition period between the first and the second phase of the illusion.

16. As far as the telescope aided viewers are concerned, once you are watching the return and it's nearly time, as I describe, very soon my advice will be only a small help at best. My viewer's guide will get you there but from that special instant onward, it's every viewer for themselves. Alone and going face to face with the big sight, the actual Big Thing is not a simple thing. It is my hope that along with the knowledge that a special sight is going to be seen created in, and out of the distance of the light of the Moon, the idea of remaining focused on not looking away, will at least be of some help to all of those who are interested and all of those who are fortunate enough to see and know the very sudden extreme degree of pressure and intensity and difficulty that is faced and actually even felt.

17. Once the object is sighted, know that a very special sight is there to be seen and as always once again somehow it's way way more than that! The ultimate staring contest is way more than just on as now all viewers are themselves realizing that they really are looking at and incredibly being noticed by the all time ancient mystery. Force yourself to absorb the pressure that the light brings. It is good for you although while you are actually feeling the light you may find yourself wondering if it's doing damage internally. The feeling of the movement caused by the light is suppose to happen as far as I understand it. It's not painful and it's not harmful. It isn't even an unpleasant feeling but certainly it is an unusual and very different sensation to feel the changed light find it's

way into the place were you are already wired to receive it. Hang in there and stay looking!

18. Along with all of the intensity and difficulties that this brings and that I know, certain things turn out to be easy. Like the fact that this happens. For me this is easy. By a total unexplainable completely lucky fluke, I saw it. Somehow I saw the big place roll by last time and I saw what the big rolling place does. I can't explain it, but I can try to describe the sights that I saw.

 I was going to add onto the above viewer's guide and finish it but instead I decided to include it and leave it unfinished just the way it is now. I can only hope that after reading the various descriptions that I am offering one day you will also be able to look back and remember your journey to the end of the Black Road. I hope that you will see the shadow's point cut left and right zig zagging and drawing the squiggly line that it does draw and actually will draw for you again when you focus and remember. If you are lucky enough to get noticed then you will be able to understand the shape effects especially if you really try hard later. Focus on the Moon's changed light when you are actually seeing the Moon's changed light.

 Try to realize that when you are suddenly feeling the pressure and indescribable intensity that this is the place that you want to be and it won't last. Push away the fear and the worry and try to think about enjoying the feeling and the sights that you will be seeing. If I'm lucky enough to be back seeing again these are the types of thoughts that I am going to try to force myself to have. I hope that this is what will happen for me, my two children, and for you and yours! It is the ancient mystery and it is the sight of the Awesome Good!
 Sincerely

8 RETURN VIEWER'S GUIDE #5

1. Shortly a very massive crater covered forward rolling moon coloured object is going to roll down across between the Earth and the Moon. The forward rolling, slowly sinking object crosses between the Earth and the Moon moving from the upper left to the lower right.

2. The object is smaller than the Moon. The object crosses at a point between the Moon and Earth at a point far enough out from the Moon towards the Earth so that the object's smaller size, relative to the size of the Moon, is able to completely obscure or eclipse the Moon.

3. At a certain point the total eclipse ends. The Moon's light emerges above the object's upper right horizon and is spread out wide open to the right and upper right of the object. I think of it as moon rise. Obviously the Moon does not actually rise instead the crossing object's surface is sinking down allowing the Moon to be seen above it's forward rolling surface and especially to the right were the Moon's light opens up and spreads out.

4. When the object is crossing between the Earth and the Moon the Sun's light that has reflected off the Moon's surface and is now traveling toward the viewer on Earth becomes transformed as it passes over the crossing object's fast moving, forward rolling, slowly sinking surface. The Moon's light suddenly does not look or behave at all like the sight of the Moon. Some sort of very fantastic effect is generated and/or caused by the crossing object's forces.

5. The object's largest surface feature as far as I know is a rounded

smoothly curved raised area. It rolls up into view over the top of the forward rolling, forward moving, slowly sinking object.

6. Suddenly instead of rolling at the viewer, the surface area of the object that precedes the top of the object's raised surface feature seems to start to tip or tilt towards the viewer. This signals the beginning of phase one of the Lasting Light event, The Valley.

7. The fantastic light blending and light changing has started. The light from the top of the object seems to stop and remain while the object itself is obviously still moving and rolling forward. The bottom area or regions of the object continue forward and down at the same time. The snail shape of the object becomes greatly exaggerated. The light from the top of the object somehow seems to stay where the actual real surface was while the bottom of the object spreads down and the entire scene seems as if it's closer to you as the object itself seems as if it's opening wide. The surface area that precedes the raised area seems as if it expands or lengthens. A very bizarre effect that is very hard to realize when you are actually seeing this effect.

Later after the shadow has been cast and the shadow has traveled across the object seconds before the shadow strikes and draws the squiggly line, suddenly the object reaches some sort of lower critical point. At this point the light from the object's upper regions is no longer being left behind and blended and or mixed and or tied into the light in the same way. At this new lower point it is as if the object has suddenly reemerged or fallen away from the Moon and is suddenly seen down lower actually turning or rolling again very similar to or possibly exactly like the way it was rolling before it started to tilt instead of roll at the point were the light blending and the first sight of the valley happened in the first place.

8. I am jumping ahead a little bit here but this is a critical point that needs to be understood concerning the timing of events that lead directly to the place in the light when the point of the big main shadow actually sculpts the form and the shape of the Awesome Good. When the spherical normal curving round surface area is suddenly visible over and past the object's large raised surface feature there are literally only seconds left before the shadow strikes! You are able to suddenly see over and past the raised area once the object itself seems to be fallen away down below in the overall scene as I try to describe. Hurry and look back to the right. If you look back to the right it will be easy for you to see the main group of obelisk shaped shadows racing away from you. Find and

focus on the point of the big main middle shadow. This big main shadow is easy to find and see. The central shadow is substantially taller and twice as wide as the two outer shadows that it travels with into the distance.

9. Going back after the full eclipse, the Moon's light is first seen again over the object's upper right horizon. The Moon's light is fantastically changed. To the right, the view is of a fantastic valley. It's as if the Moon's light is tipped away and spread wide and deep. You see an expansive valley with no real discernible valley floor at this point. Instead of any sort of actual valley floor at this early point before the wave of the shadow has crossed over the object, a flat white cloud looking effect is seen every where in the area of the valley between the mountains. The entire scene is surrounded around it's perimeter with the Moon's seemingly piled up surface light forming the sight of very uniform looking steep faced mountains on all three sides. A normal looking view off into the distance of a valley in a land area is seen. Even the flat white clouds look natural. The view of the valley is oriented exactly the way you would expect a view of a valley to look. A very convincing valley look is seen as if some sort of new place. This is the sight that you will see to the right of the object once the Moon has seemingly risen. It is the Moon but this sight does not actually look at all like the Moon.

10. The rounded raised feature continues rolling forward and slowly sinking. From the viewer's perspective the object's raised rounded main surface feature starts to develop into a rounded steep cliff. Gradually the smooth vertical cliff becomes an overhang.

11. Suddenly a very noticeable large black spot appears under the newly formed overhang. The Sun is located above and behind the viewer. It begins to cast a shadow under the newly developed overhang. The black spot of the shadow grows. The shadow spreads out and thickens while it seems to remain clinging to the underside of the forward rolling overhang.

12. The shadow is about to drop to down to the ground below, the surface of the object. At first the shadow seems to slowly stretch or reach gradually spreading down the area under the overhang as it thickens even more..

13. Suddenly the shadow then seems to leap or pounce down to the

ground below. The shadow then immediately very rapidly moves from the left end of the valley towards the right. The shadow is also moving away from the viewer on an angle that is always gradually increasing relative to the object and the viewer. This is the time when the shadow is young. This is when you can see the wave of the shadow crossing the actual surface of the object as it passes through and over the object's craters and the towering crater rim pieces. This is truly an awesome sight all by itself!

14. The object frames the left side of the view of the valley.

15. The wave of the shadow moves beyond the object's far upper right horizon and into the distance of the light from the Moon to the right.

16. Below in the scene on the left, the object's forward moving forward rolling slowly sinking surface has moved lower relative to it's original position when the Moon the valley, was first seen. The light from the object's surface area that bordered with the Moon's light at that earlier point is somehow still there. The rolling surface of the object connects or ties into the the Moon light the valley itself through the area of light from the object's surface that remains behind still there somehow. The effect for the viewer is one were even though the object's surface moves and rolls and looks very different from the valley everything you see from the left where the object is located all the way to the far right in the unmoving valley seems to flow naturally from one to another. A very easy and natural transition between the moving object and the unmoving valley is seen. Everything from the left to the right seems as though it is one expansive place. The overall effect is a very natural looking and very convincing valley scene that extends from the great distance of the view to the right all the way back to the object on the left.

17. During the time that the shadow's great wave travels a group of big main shadows start to separate themselves out. The points of this group of long narrow shadows move away from the viewer towards the distance of the valley to the upper right further and faster than the leading edge of the shadow's fast moving big black wave.

18. From nearly directly below your eye's very far forward viewing position three big main shadows emerge from the wide group of now very tall and long narrow individual shadows that have emerged from the shadow's wave.

19. All viewer's need to ultimately locate the big main center shadow that leads the main group of three shadows

20. Within the three main shadows the middle shadow is twice as wide as the two outer shadows.

21. The middle shadow is taller and longer than the two outer smaller shadows.

22. This central very big main shadow is now very rapidly moving into the distance of the valley's emerging newly created floor.

23. The whole idea is to focus on the obelisk shaped point of this now very big black shadow that resembles a very rapidly moving growing black road.

24. The size and shape of the big main pointy black shadow grows all out of proportion when compared to the size of the object.

25. Defiantly all viewer's now want to be very focused on the big main shadow's point basically as soon as possible.

26. Once the wave of the shadow has left the surface of the object all viewer's do want to look at the object's far sloping surface's brilliant and very fantastic shapes that remain above the height of the wave of the shadow. It's a must! There is time for a look at these features that are the light from the tops of standing crater rim pieces that are higher than the height of the thickness of the shadow. A very fantastic sight that along with the rest of it needs to be seen to be believed. There is only a very few short seconds available to the viewer for viewing the object's far slope.

It may be that as the object's surface continues moving across and down the light from previously obscured stars and maybe even planets may suddenly be seen in the area down over the far slope along with the light from the tops of crater rim pieces. In the area of the far slope initially or suddenly later I am not sure which two large triangle shapes are seem.

These two very spectacular attention grabbing sights will instantly be very noticeable and easily discernible to viewer's who are looking in this region once they appear. The triangles are located in the foreground in this region. It could be that these triangle shapes are set back within this region but because of their large size and their extra brilliance I may just remember these shapes as being set forward towards the viewer.

I saw these triangle shapes either just before I looked to the left and then saw over the object's raised feature at the normal shaped surface area beyond or I saw the triangle shapes just after I saw over the large raised feature while I was looking back to the right only seconds before the shadow reaches the ancient place in the light were it sculpts and creates the form of the Awesome Good. My best guess is that these two large triangle shapes are not visible initially once the wave of the shadow has passed over the object's upper right downward sloping horizon, but instead slowly emerge or suddenly appear in this region at a later point. I saw the many shinning shapes that are visible above the thickness of the shadow down over the far slope right after the wave of the shadow had passed off the object and I did not notice the triangle shapes. Perhaps the triangle shapes were actually visible at this early point but they had not grown in size and brilliance and prominence yet.

I looked left and right and up and down a number of times so I did not see all of the sights that present themselves and I did not see the every detail or sight that evolves throughout it's total evolution. I have to guess and try to piece things together the best I can. Basically I have the above sequence of events narrowed down as I describe. I am not completely 100% certain in every area as I try to explain.

I was not watching the last return and at the same time taking notes and trying to remember exact details and the timing of the sequence of events somehow knowing that thirty plus years later I wound be trying to recall every exact specific detail along with the timing associated with the sights and events that are seen so that I could then write it all down in a coherent manner with complete 100% accuracy. That sort of task would be very difficult or even impossible for even the most prepared viewer who was even ready to put pen to paper immediately after their view of a return was over never mind a kid who was taken completely by surprise and sent home stunned and reeling not having any sense of the importance of the sights just witnessed. Consider my circumstances and it should be no surprise that I am not completely 100% certain regarding every detail and the exact order and the exact timing of events. The main points I mention are very accurate.

I understand that future viewers will be easily able to follow the details and the sights and the order of events as I describe them. For me this is easy to understand and I am not at all concerned in regards to the main points I mention in my viewer guides and on my website this web site and in this book.

In an effort to be as specific as possible and in an effort to raise other certain points for consideration I have decided to try to expand and discuss possibilities in areas that are in areas that are not 100% clear to me. I know of many points that I could mention similar to the points I am mentioning in regards to the scene down over the object's upper right horizon above the far slope. I am trying to narrow things down so that I can try to mention important points and sights so that future viewers can have a wider understanding if possible. This is especially true in regards to trying to provide future viewer's with indicators that will be a signal to them concerning the amount of time left before the big shadow's point draws and sculpts the sight of the Awesome Good at the end of the Black Road in and out of the changed light of the Moon.

This is a general viewer's guide. I understand that a viewer of this event could easily get lost in the details that are seen. Many details are seen that I could not ever hope to adequately describe in writing. As always there is always much much more as telescope aided future viewers will be able to clearly see for themselves when they are actually seeing this sea of details for themselves. I try to point towards the biggest main points and I try to touch on other notable points. Overall later after the next Return people will see and clearly understand that I am only skimming the surface concerning all the sights and all the details that are seen. I am trying to actually complete pages so that I can actually start the process of putting these web pages out there so that people will have heard the idea of what happens during a Return.

Not only is this a starting point for interested persons this is a starting point for me as well. I have never been able to talk about, write about or think about all of this in one sitting or in just one go. Over time I will continue to chip away adding more and trying to build a bigger clearer picture as I go. I am trying to be accurate and helpful and at the same time I'm trying not to get bogged down spinning my wheels in the details.

These points I am mentioning relate to the fact that I realize that overall there is a very good chance that many people who are viewing a return especially from up close with the use of a small backyard telescope may well find themselves lost in the details. It is obvious to me that I could have easily been looking the wrong way at the wrong things at the all important instant in time when the point of the big main shadow was sculpting the shape and form of the Awesome Good.

I could have missed it even though I was watching the Return. That's actually what very nearly happened to me. I happened to look left in time to see over the raised area of the object itself and then by complete sheer luck I looked back all the way to the right again and saw the big main shadow traveling away at a point in time approximately not even more than three seconds or so before the point of the big main shadow went up and down and up and down. In the next split second the shadow's point draws the squiggly line as I describe. It is only by sheer luck that I happen to know that you see over the object's raised area on the left seconds before the shadow strikes on the right. I was totally lost in the details during the time before the shadow suddenly draws the squiggly line. Just by luck I happen to know what happens on the left before the shadow strikes on the right.

Certainly if I had of looked left for even a few seconds longer I would have no idea that looking at the big main shadow's point is the overall big main most important point of it all. If my eye was not looking at the very point of the big main shadow would I have found myself looking down from a place just above and behind the sight of the back of the Awesome Good? I don't know maybe. I just don't know because that's not what happened to me. The way I saw in the light and what happened to me is all I have to go on and the only thing that I know with 100% certainty.

As I describe else where I know that a viewer can join in at a later point and actually see because of what happened to my friend's dad who also saw. Certainly it's as simple as actually looking but having seen the sight of the back of the Awesome Good from above and behind I clearly understand that this place that you see down from is definitely the place were all future viewer's do want to be plain and simple. All of these separate sights I describe happen quickly and then they are over. My entire view of the Return lasted less than a minute maybe thirty to forty five seconds from start to finish. Judging the entire length of my view does get into an area that turns into a bit of a blur for me. I am certain that my view was less than one minute I think and at least thirty seconds. Judging time is a very difficult thing for me to do with accuracy as I mention elsewhere.

All the separate sights I describe take only handfuls of seconds or split seconds. You are going to see a very fast paced series of sights and events. Everything happens very fast and a viewer really does only have one chance to get it right. You don't want to be looking the wrong way at something else when it's finally time to look down from above and behind. All viewers do want to get noticed! This is why I put such a high priority on the sequence and the timing of the sequence of events. Once people see a return and they have seen the changed light for themselves, they will be able to look back and understand why I have spent so much time going over and over the bit about finding and focusing on the point of the big main shadow and the part about what you will see happening if you are looking the wrong way seconds before the big main shadow strikes in the opposite direction. They will have looked down and seen the back of the Awesome Good and if that's were they were seeing down from in the changed light then they would have been there suddenly seeing the Awesome Good turning and getting up to his left and they would have been there for the Big Noticing! This is a central and a very important part of a return and what happens in the changed light from the Moon. I am very fortunate to be able to help direct future viewers to this all important place in time in the Moon's changed light.

I hope that many many eyes will all be seeing down from that place in the light the next time it's time to get noticed and go face to face with the Man of Light. Overall if that's the way things happen the next time then I will know that I have made a difference. I will have done the right thing with what I saw and what I know. Some things worry me greatly and keep me up at night. The points I mention in my viewer guides are definitely not among the things that worry me or the things that cause me to lose sleep.

The ancient things that I have been fortunate enough to find and read although sometimes or rather most times jumbled up do mention many specific points that I can understand and many specific points that do clearly point to what I consider to be the clear and obvious fact that as far as the parts of a return that I saw are concerned everything happens the exact same way every time. Perhaps the final evolution of the entire event all the way through to the end of the world of light brought forth, the oval shaped   plain may change. In fact this does look like it could possibly be the case however as far as the initial sights and the initial sequence of events is concerned I definitely have the feeling that the specific sights I saw repeat in exact or in near exact detail. My descriptions of the sights I describe turn out to be accurate and reliable. I know that future viewers may even be surprised at just how accurate my descriptions really are once they have seen the sights I describe.

27. Once the object's normal surface area is suddenly visible once again on the left over the top of the object's raised surface feature the overhang time has nearly run out.

28. The overhang continues to roll forward and very slowly sink. After the wave of the shadow has left the object's surface the object's surface area that is beyond the top of the overhang becomes visible. The object's original spherical curving shape is seen again.

29. If the viewer sees the object's normal looking surface area over and beyond the overhang there are only a few seconds left!

30. All viewer's now need to very quickly look to the right. Find and focus on the point of the big main obelisk shaped shadow as soon as possible!

31. Find the big main point and focus on not looking away at all any more. The sight seeing is over and the big main event is about to begin!

32. There is no chance that anyone will actually be able to prepare for the truly awesome shocking sights that are about to be seen. Even though there is little advice that could possible help very much at all I will try to give some simple advice that I understand to be advice that is definitely extremely basic and extremely important.

33. Simply said, look and keep looking. This is something that somehow is not an easy thing to do. The viewer will be facing an intensity that is beyond belief! I understand that there is simply no way that I can possibly convey to others any sort of understanding concerning the very incredible intensity and pressure that you will really feel. Knowing ahead of time that the intense pressure and the feeling that you will feel is not harmful as far as I understand it may help. I resisted the flow of the pressure or at least I think I resisted the pressure. I don't think that my reaction helped my situation. I just simply had no clue that I could feel anything through my eye all the way in through to the place in my head where light goes if you are seeing it.

Certainly I know I was instantly very surprised to know that I was suddenly feeling anything at all never mind feeling the way the light forces itself into you. The feeling of the pressure of the light is simply even incomprehensible. It turns out that light has some sort of pressure or at least this very special changed lasting light has some sort of mysterious and very real pressure or force behind it or with it when you see it and when you see what it suddenly looks like and what it does!

34. This has always been a problem for me to understand. I know this but I have no clue as to how something like this works. I saw light looking solid as if a statue and moving much slower than I think that light is suppose to move. Certainly I had no idea at the time that I was even looking at actual light when this was happening. I only figured out that I was looking at light, much later.

35. Does anyone know that light can do these sorts of bizarre things? I have no real true idea, maybe but on the other hand I have a feeling that maybe this is not 100% completely understood. I do know for sure that very soon all the those big scientist types are going to see and feel the way light moves as if a statue over there going fast but actually going much slower than light is suppose to go. Do they know this already or is this sort of sight going to be a bit of a shock? I have a funny feeling that even Mr. Einstein would have been sent reeling scurrying back to the drawing board! We shall see! I think that soon it will be time to get out the chalk ... Then soon after that all you scientific type folks can reboot your computers with all the new things you suddenly were forced to know! Who knows maybe the way the changed and lasting light behaves was predicted by Einstein and once again his theories are proved by observation.

36. Multiply these sorts of surprising types of difficulties with the impossible way that the light actually appears and the way it looks at you face to face and suddenly you have arrived at a place that is it's own extreme shocking reality that grabs you and forces you to know and realize what you are seeing! Impossible contradictions that are real and staring at you overwhelm and constantly overpower you again and again with every new stunning sight! I really hope that one day I get to sit around with a bunch of other people who have also seen these exact same sights. I am very curious to know what sorts of opinions and experiences and feelings will be offered up and going around the room. Suddenly I won't feel alone anymore. I will feel much better.

37. The entire point is to see. All viewer's will need to continually refocus on the idea of not looking away so that they will be able to

remain actually looking. Brace yourself and try your best. If you don't keep looking, you will regret it later.

38. Focus on the very point of the big main shadow. Allow your eye to follow and ride this most distant and central main shadow's point's movement and you will find your eye's viewing position to be magically drawn always farther forward into the distance of the light of the Moon.

39. Very suddenly the point starts to very rapidly go up and down and up and down. At this point the big main shadow reminded me of a flat ribbon as it briefly rapidly when up and down and up and down.

40. Almost instantly or instantly after the shadow's up and down motion stops the cutting left and right zig zagging squiggly line starts to happen!

41. If you are looking you have arrived! The intensity is yet to strike but down below and in front of were you are seeing down from the sight of the very muscular back of the Awesome Good is there!

42. Your eye's viewing position is from just above and behind. It is as though you are looking down from a height of about 15 - 20 ft. Maybe slightly higher but from no more than 25 ft. The size and power of the telescope being used may be a factor to consider here. A second or two or maybe three and then suddenly the ultimate instant in time is about to happen next! But just before the ultimate instant happens the lasting light moves up. The sight of the back of the Awesome Good drifts up closer when the light moves up. Suddenly a flat smooth place or area is seen surrounding the Awesome Good.

43. In between the light's upward motion the light somehow magically builds outwards turning the flat white cloudy effect that surrounded the Awesome Good into what looks like flat hard ground. You can see right out to the edge of the flat smooth area. Beyond the edge of the flat smooth floor or ground area where the Awesome Good is now seen positioned you are able to see the indiscernible look of the flat white cloud effect again. This is true out to the left of were the Awesome Good is positioned. I did look out to the edge of this sudden new flat area to the left of the Awesome Good and so I know that there is an edge to be seen in this one spot. I don't know if there was any sort of discernible edge to be seen anywhere else radiating out from the Awesome Good's central position. Flat ground did surround the the spot were the Awesome Good is situated at this point and if you look left you will see the edge of the

new flat smooth area and past to the flat white cloud effect.

44. The changed and now sculpted incredible solid lasting light from the Moon in the form of the back of the Awesome Good is very clearly there! The viewer has only enough time to realize what they are seeing before the impossible Awesome Good that they have suddenly found themselves looking at suddenly gets up and turns to the left all in one fast spinning motion and turns to face you and look back up directly at you!

45. It turns out there really is an ancient Big Thing and this is it! Instantly at this same incredible instant the beginning of the completely overpowering mind bending intensity is nearing and building and getting ready to strike! An incredible thing is happening! The way this light forces itself is at the heart of the question of human life itself. If you are looking you are made to see and comprehend. The shape of the light that has focused on you must have been first. Humans are here on Earth! How does this work? How could a situation like this exist? I certainly don't know but somehow this situation does exist.

46. In this now completely impossible situation the simple advice I offered concerning the idea of remembering to try to focus on the idea of not looking away will be at best, only a very very small help. Remembering that you are suppose to look and you are suppose to look for as long as possible is a good place to start.

47. Also it is plainly obvious to me that the changed light from the Moon is beneficial and good for us. It must be. I don't know how it does what it does but this fantastic changed light does definitely have very special qualities.

48. The view of the valley phase of the changed light event ends with the place where the Man of Light is positioned gently moves up towards you on a gentle slope twice while remaining face to face with you. At this point your viewing position is still located within the flat white misty area of the Valley floor. Now very suddenly a short transition period takes place before phase two of the Changed Light event the rising Lasting Light Mountain begins.

49. During this transition time period the view of the valley transforms downward in in size. It is still there but scaled down or focused inwards. The first phase of the lasting light event the valley is over. The transition period with it's repeating inward pulses ends.

50. Now the second phase of the lasting light event has started. The light lasts and rises as more of it arrives from over the object's upper right horizon. The light from the Moon arrives from over the object's upper right horizon as one way arriving and slowed down areas or waves of basically moving as one areas of almost solid moving light. Incredibly viewers are going to literally see rising and lasting real light from the real rock of the Moon.

A very convincing looking mountain is very suddenly rising up through space growing at it's base complete with the valley contained down within the upper structure that is different from the bending curving column of mountain that has formed below. Once the one way moving areas of light have joined in at the base that is the point when the light, that is already effected and changed actually becomes solid and lasting.

51. A completely incredible totally stunning and because of the intense pressure that is felt an even overpowering sight! Concerning the upper structure itself I have seen pictures of ancient Maya buildings that are exact copy's of the way that the upper structure looks. Certainly the entire bending curvingly straight height of the mountain of light is not constructed but I think that the mountain's upper structure itself is the focus of the builders. Definitely this entire event was very much the focus of the ancient Maya as well the rest of the people of the past. In fact one day it will not just be me but everyone is going to realize that a great majority of ancient constructions are in fact models of the way the lasting light looks. I know that this can only sound impossible but if you have ever seen the lasting light you will have been forced to know that this is simply the way it is. It turns out that as usual it's almost always about the lasting light and the way it looks at you the way it appears and the very bizarre strange way it behaves.

52. The lasting light mountain is rapidly rising and the face to face ultimate staring contest is still on and has not been interrupted by the transition period's inward pulsing or by the way the actual Awesome Good itself has changed or evolved in appearance during the transition period. Now at this point the full force and intense pressure is instantly suddenly being felt and literally received by the viewer. I realize this can only sound impossible I agree. Anyways I can't let what should be impossible be any sort of factor for me in my decision to try to describe my view in an effort to provide future viewer's accurate information, although it's not easy to ignore these sorts of problems that I face.

The impossible is going to be happening again and I realize that I obviously can't change or effect how the light looks. What am I suppose to do? Not describe what the lasting light looks like because of what and how people are going to think? At one time I used to beat around the bush when I wrote, now I don't. It is what it is and the light does what it does. It does not matter how impossible this sounds to people. It turns out that the opinions that will be expressed against my descriptions are completely totally irrelevant. I don't mean to sound rude towards people and it's not as if I don't respect the opinions of others, in fact I do have a tremendous amount of respect for the opinions of others. As it happens this has nothing to do with what I think or what anyone else thinks and I am able to realize this fact. If a person can't stop to hear the idea of a return then that's fine. I'm not trying to convince anyone, that is not a thing that is possible for me to do. If you want to know a bit about what the changed light looks like then read on. If you don't want to know don't read any more of my descriptions. This is difficult to realize. Plus why should you believe anything I say or write? This viewer's guide is for interested persons. I am not try to actually get you or anyone to believe my descriptions. I am offering you a chance to read and hear about what happens during a return. You don't need to actually believe any of this. Sorry once in a while I seem to have to go through this sort of spiel. I suppose it's in defense of myself. This is part of the difficulty for me trying to describe my view. I always cringe to it's not just you. I saw it and I can't believe it either.

Anyways it has literally taken me years to turn back around to face this again. This is to big for me and I pretend to have found my courage but I really haven't. Anyways here I go again I don't want to argue or debate. I'm simply trying to provide information. In the end this is a for your information thing. I did see the big thing that happens between the Earth and Moon and I am trying my best to describe these particular very strange and totally bizarre sights. I am doing this because it's the right thing to do and because I love my kids and I want my children to grow old and I want their children to grow old. I want everyone's children to grow old. The sights I describe and the points I mention are accurate and genuine. It turns out that this is not easy and it turns out that this is just the way it is.

53. As each wave or area of moving light arrives from over the object's upper right horizon and then moves and forms the next area and solid moving light it has joined in the base of the mountain and is itself the newest area of mountain rising. Once the areas of one way moving

pulsing forward waves of light have arrived and have finished moving as it's own area of moving light and has become the newest area of rising mountain they remain unchanging. It's the light from the rock of the Moon. You are seeing it but it's this great big huge thing rising over there. A very very convincing mountain is seen although you don't normally see mountains rising bending curving up through space growing at their base. It is what it is and it does look like rock or more like smooth stone.

In fact as a kid for some years I thought of it as a smooth sided stone mountain not a rock mountain which I suppose I was thinking would look rougher. As far as the very strange waves that arrive at the base and become new mountain rising I was not really able to think about this very much. I knew the mountain was somehow magical but for a long time I had no idea that it is actually light. Even once I finally figured out that the telescope had not been moved I still had no idea that the magic stone mountain was light from the Moon.

54. What happens when you are actually looking at something say a tree? For me I don't automatically think to myself ah ha there's light from a tree. Instead I simply see a tree. That's what it is it's a tree not light. If you see the Magic Light Mountain that's what it is a mountain. A strange mountain but you see it the way that it is and that's what it looks like. When I suddenly saw the rapidly rising Magic Light Mountain that's what I saw, a mountain. I didn't realize anything more. This was a very strange mountain but it was a mountain. Even though it does this bizarre thing and it has the Awesome Good along with the other forms sculpted by shadows located down within the upper structure with the Awesome Good it looks like a familiar thing. It looks like a mountain that is as solid as stone.

It turns out that in many ways the mountain is light from solid stone or at least from something solid the Moon. The Moon is real. When you are looking at the changed light mountain you are looking at real light from a real place. In spite of the bizarre nature of the changed light event it turns out that in the area concerning the mountain's appearance it should be no surprise that the mountain itself looks real. It does look like stone or rock as some others in the past have also described the mountain.

This is an unbelievable thing I know but I understand that future viewer's will benefit if they have some idea what they are seeing when they are seeing it. The fantastic sights that are seen above the object's forward rolling surface are all made of light. It's all real light from real places mostly the Moon but perhaps a number of stars as well and maybe even planets I don't know but on the most basic overall fundamental level it turns out that the vast majority of all of the incredible sights that the viewer sees above the rolling surface is light from the Moon after it has been somehow changed by mysterious power of the big rolling place that returns and crosses between the Earth and the Moon.

55. Certainly this does sound impossible but it turns out that this is what really happens. Ultimately a situation is developing and evolving were below you can clearly see the turning rolling crossing slowly sinking disc shaped spherical object. Above the object the lasting light mountain, tall and narrow curves to the left back over and against the object's left to right motion and then ultimately the oval shaped plain the surface of the Moon complete with the forms sculpted by the points of all the shadows emerges out from down inside the upper structure of the lasting light mountain. At least, up to whatever exact point in history the oval shaped area of lasting light emerged from down inside the top of the mountain of lasting light. This is exactly what use to happen and that's that as they say.

56. Then mu guess is that the formerly curving straight light mountain turns into a snake like line and the object separates from in front of the Moon falling away down further towards the lower right. Check "The Great Serpent Mound," located in Adams County, Ohio, U.S.A. Ultimately that's the scene. My view ended while the light was still building at it's base. I did not see the oval shaped light from the surface of the Moon emerge from down inside the lasting light mountain's upper structure.

I did not see the eclipse actually end. The Serpent mound shows the changed light no longer emanating from over the object with the Moon partially behind it the way it was doing so during the eclipse. After the separation the light has not finished emanating from over the object. The spiral effect is the rolling motion of the object with the light still emanating from over it's surface. Another guess.

57. The area that was originally the flat cloudy area within the valley's Mountains becomes the plain or the firmament the oval shaped area. The mountains that were originally seen surrounding the flat white cloud of the valley became focused inwards and became the walls of the sunken room. Later these same mountains become a sort of sleeve of lasting

light. The outer walls or area of the rising lasting light mountain. Later after the oval shaped area of changed lasting Moon light emerges out from down inside the tall narrow bending curvingly straight mountain, the mountain itself the sleeve of outer light that contained and stored the oval shaped area of lasting light is straight no more and it falls away down to the right or it just becomes a squiggly curving snake like line. Then the oval shaped plain continues towards the Earth and is described in ancient texts and it the focus of the ancient texts.

58. The original face to face staring contest starts when the Awesome Good turns around to the left and stands up to directly face you. During the rising of the growing mountain you are seeing from a position that is back much closer to a place that is much more like what you would expect a normal viewing position or perspective to be like. From this much more normal position there is much more distance between your position and the position of the rapidly rising nearing position of the Awesome Good compared to the face to face moment. This is true but at the same time with the sudden pressure and screaming intensity that is suddenly felt and known by the viewer the ultimate staring contest is on even more than ever. While rising the impossible Awesome Good is still there going face to face with you from his central and forward prominent position.

59. Suddenly it is clearly obvious that the focal point of the changed light is still the focal point and it is still defiantly squarely focused on you but it looks different. It has changed! It's the second version of the ancient Man of Light. He transformed during the light's inward pulsing transition time period and it now has a whole new look although his eye's focus has retained it's original penetrating stare! An incredible thing to see!

60. As the light rises the object's upper right horizon looks very much like a very thick lid being drawn slowly back open showing more as the sunken room rises. The valley area is seen reduced in size now a sunken room now down inside the top of the mountain as the mountain continues rising bending curving very directly very rapidly directly at the viewer.

61. Down inside the top of the rapidly rising mountain of changed Moon light the Awesome Good is closing towards you. Your eye's viewing position was returned back to what you would think of as being a more normal and far back viewing position during the transition time period between the valley and the beginning of the rising mountain.

62. Incredibly the telescope added viewer is going to see that this intensely focused light is rapidly nearing. Incredibly down inside all of this this now huge growing thing the Awesome Good the focal point of the entire event itself with great speed and momentum is very rapidly closing on your position!

63. Down inside the mountain's upper structure the valley now looks exactly like a sunken room. There is an array of separate human forms located around the focal point. All of these separate forms or main elements are created by the other main shadows that reached further faster than the shadow's main wave but after the main shadow sculpted the form of the Awesome Good. Just before my view ended I looked away from the Awesome Good and to the left and slightly up around the curve of the shape of the light from the curve of the object's upper right horizon at the next human form positioned to his right.

64. The next form over in the direction I describe may have been created by the shadow that traveled to the left and slightly behind the big main shadow or it could have been created by the shadow that was the next main shadow further to the left I'm not sure which. I am guessing but for various reasons I am leaning towards the idea that the next main shadow further to the left is responsible for sculpting this form that I am referring too. I looked at the top of this form and then down the front of this form to the valley floor below now seen as the floor of the sunken room.

65. My eye went from the bottom of the front of this form across the floor traveling left in the view finder to the base of the sunken room's wall and then up and out of the inside of the sunken room to the outside, outer wall of the sunken room which is actually the outside of the top of the rising mountain, the upper structure. Again, my eye traveled from the front of the base of this second form that was created in the valley to the left towards rooms wall.

What is seen as the walls of the sunken room are actually the steep faced mountains that were originally seen ringing the valley. The sight of the sunken room is actually the valley rising reduced in scale with the forms that were created by the shadows, contained and arrayed within the floor area of the sunken room. A very incredible thing to see.

66. Once my eye was outside I saw how the upper structure looked and I saw how the rest of the mountain looked as it curved back down to it's base. I looked all the way back down down to the mountain's base from

the edge of the sunken room, the top of the. light mountain then I looked at the object's forward rolling surface. The last few seconds of my view were spent watching the newly arriving areas of slowed down and moving areas of light smoothly and naturally joining in and becoming the base of the mountain as new areas of created and unchanging areas of moving rising mountain. These one way areas of moving light pulsed forward and joined in as new mountain with a pulsing motion or movement that is very similar to the way the the transition period's inward focusing motion pulsed inwards. I can guarantee that future viewer's will be totally amazed by even just the sight of the bottom of the mountain building alone. All by itself the the sight of the way that the light arrives and joins in with everything that is already there, creating and storing more layers of light is truly an fantastic spectacle!

67. This action is actually central to the entire lasting light event. These new pulsing one way waves that arrive and then join into the base of the mountain are like the next new frames in a long stripe of movie film footage. Each forward pulse is it's own layer in the mountain and another picture in a continuous movie to be seen later as a part of and within the plain. Also each separate pulse or wave or layer in the mountain records and stores a time in the light with the continuously evolving influences of the shadows cast off the moving rolling object's crater rim pieces into the light from the Moon. Each one way pulsing area of light is a type movie onto itself.

68. The entire scene on the plain and all the forms that are seen on the plain are seen in action later by viewers on Earth as if they were a part of a fantastic movie. The light records and stores transports and then plays or replays all of the results of the shadows sculpting and creating in the light that happened earlier when the shadows were actually being cast off the object into the changed light from the Moon. As the one way moving areas or ways emerge from over the object they contain everything that just happened and they contain everything that is seen happening later.

69. As usual this is to bizarre to explain in words or even simply to bizarre to understand at all. I am able to read ancient texts that describe the way the light arrives and forms or reforms or is presented when the plain emerges and is presented. I am able to read on through these descriptions of what the lasting light does next. I saw the first part of the sculpting in the light and I am able to read about what happens next starting with chapter one. This is truly a stunning overwhelming thing to realize. I am always in some sort of shock because of what I saw and as I

realize more I am always only more overwhelmed. Like I said, it's plainly obvious to me that I saw the big thing that happens. Also it's plainly obvious to me that this big thing that happens is about to happen again very shortly. I know this is impossible for future viewers as well as for everyone else but BE READY!

70. I don't know what be ready means but that's all I can think of to write at this point. I know we have to be ready but how are we suppose to actually do that? We need to see the object returning before it gets here but is it possible to actually be ready? I don't know but in reality we are not going to be ready. These types of questions and the answers to these types of questions are way to difficult for me. I am able to realize that these sorts of questions need to be asked and they need to be answered. This is obvious to me.

71. Then unfortunately my view ended. I describe and discuss this point in detail elsewhere.

72. Will the entire evolving scene remain within the telescope's view finder right through the entire event including phase three of the changed and lasting light event? I don't know. I think that having binoculars available may prove to be a good idea. Along with possibly enabling the viewer to step back a bit in order to get back from the big sight, binoculars may provide a degree of aiming flexibility that may or may not be required.

73. I think that part of the point that I am trying to make at this point is that viewer's should try to be ready for the next sudden unexpected evolution of the lasting light event. Later during phase three of the lasting light event the oval shaped plain the light reorients itself somehow and this causes your eye's viewing perspective to change. From what I have seen from looking at ancient depictions the viewing perspective or angle that the viewer has when looking at the valley is the same or very similar to the angle that the viewer has when they see phase three of the Changed Light event the plain.

74. First your eye's viewing position moves forward drawn closer. Then your eye's viewing perspective is from farther back as you watch the mountain rise. Later during the third phase or the oval shaped plain when it appears as if the original but transformed or evolved area of the valley floor is seen again minus the original surrounding mountains from up close it may be possible to see the entire event through a telescope

without the need to switch to the lower magnification and wider field of view that would be provided by binoculars. On the other hand maybe not. This gets into unknown territory for me. Maybe the naked eye will be all that's required just like in ancient times. I think that binoculars may be the best practical choice. Telescopes are tricky to aim with the constant adjusting that is required for your typical small backyard telescope due to the earth's rotation.

75. If I saw it I know it and I can describe it. As far as the rest of the visual event is concerned I have to guess based on what I read and what I see in ancient depictions. It looks very much like a small backyard telescope will work just fine right up to the end of phase two of the illusion, the rising mountain of light, or nearly up to the end of the second phase of the illusion. Certainly the view through a small back yard telescope would have worked out just fine for some time past the end of my view that night long ago, exactly how far past the end of my view I can really only guess. Definitely in ancient times when the third phase of the illusion occurred and the Moon's changed light arrived at whatever height above the Earth, a telescope was simply not required and obviously was not available. Also if the oval shaped area of changed light actually arrives here near the earth a telescope may become completely useless. Binoculars may prove to be worthwhile along with the naked eye at this point. In time this question may be answered.

76. It turns out that it's all about the Moon's changed light from the start to the spectacular finish. Will the entire lasting light event occur and play out in it's full entirety? Will we see the world of light arrive and then age and then end? Will we see the end of the world of lasting light just like the ancient people saw and describe? Even if we don't see the sight of the lasting light reach all the way to the end of it's fullest longest potential time period along with the entire event from start to finish like it used to happen in ancient times everything we will see will be spectacular!

77. We have a chance to see and know what our most distant ancestors saw and knew. It turns out that in some ways, in some areas, our distant ancestors did have a better understanding about what goes on around us in our world and in our own local area of space. They knew this fantastic thing and we don't. I am sure that our ancestors never could have imagined that knowledge of something so big and so fundamental to everything could somehow one day be basically lost. Listen to how they wrote.

The audience of the day simply knew what goes on plain and simple. It seems the authors of the day never stopped and turned around and started at square one in an effort to take it from the top for the benefit of those who did not know. Everyone knew and it's obvious that many ancient people did have a deep and very sophisticated knowledge and understanding concerning the object and the details within the lasting light and for whatever reason we don't.

78. It turns out that there is a speeding massive standing stone crater covered forward rolling moon coloured incredible celestial object returning and it's just about back. Ready or not! There is no stopping this from happening. This is going to happen again as always perhaps soon maybe May 26th 2018.

79. Fortunately whatever the down side if there is a down side the sight of the ancient mystery does happen over the object's upper right horizon in the fantastically changed light from the Moon. Incredibly it's all about light that has been changed by the forces and raw power of this truly fantastic object!

80. I say fortunately. because I am glad that there really is a big thing that happens. The Ancient Object can cross down in front of the Moon safely and is a very fantastic thing plain and simple! It turns out that the ancient texts describe an event that really happens. Many people are convinced that the ancient texts they follow are based on real events. This is what I see around me and this is what I hear around me. It turns out that the ancient texts actually really are based on a real event. This is a fantastic thing at least in my opinion this is a fantastic thing. People are all entitled to their own opinion and so am I.

 The ancient object itself is beyond spectacular and very incredible! The way the object's powerful forces change the light from the Moon is simply the most fantastic incredible thing that happens anywhere, anytime! Good Luck and remember there is such a thing as a safe Return. The Ancient Object's last return orbit down was a safe return!
All the best to you and yours,
Thanks... D.S.W.***

9 THE FUTURE
AFTER THE OBJECTS NEXT RETURN

A main most important point that needs to be made is that the Object can return and cross down between the Moon and the Earth safely. This fundamental basic fact will not be readily apparent to people when the Return starts happening but this fact is extremely important. Once the returning Object is observed this fact will be understood because observers will see that my descriptions of the Object will match. Knowing the Object can orbit across between the Moon and the Earth safely is a basic and very important starting point for people on that difficult night that's fast approaching.

There will be those who will say that the return of the Object of the Crossing Down automatically equals doom and gloom and destruction on the Earth or even the destruction of the Earth itself as a result of some sort of collision between the Object and the Earth. Clearly obviously the returning orbiting Object never hits the Earth. The people who will be saying this are basing this doom and gloom narrative on what exactly? Best guesses? We all have to start somewhere. I make guesses as well and I know I won't be right all of the time either. Basically we do our best to guess based on the best information we are able to gather. If you are a scientist you'll call it your theory. I base my guesses on what I know from what I saw and I know for myself with certainty without having to guess that the Object will orbit across and down through between the Earth and the Moon again as usual and then it will be gone again. At that point it will be starting it's very long journey back down home to that special place between the Earth and the Moon.

The Object that I saw is in a very precise orbit that sees it cross down through between the Moon and the Earth. The last two times the Object crossed down, 1926 and 1972 it seems that overall at least basically the people of the Earth did not notice. For me personally this is reassuring evidence that I can have hope for and expect to see another safe Return just like the Return I saw. I know there's such a thing as a safe Return. Based on this I am looking forward the future after the ancient Object's next Return. I know there are no guarantees that the Object has always and especially will always orbit down safely. In fact this does point towards what are my worst fears for the future of my family and my children and the entire world's children. The flip side of this is I know there are real reasons for hope for a real future for my children, your children and the children of the future.

Everyone says the world is one messed up place. Yes it is as we all know and it's humanities' fault. We are ruining the planet and we can't get along with each other. I think the facts of what happens during the Object's crossing and specifically the knowledge of the Changed and Lasting Light is humanities' best hope for the future.

A big rethink is nearly upon us. We argue over and even kill each other over the details of what happens in the Changed Light. Or rather should I say that we argue over this or that line of text that describe the Lasting Light sights and what happens in the Changed Light. Many times ancient people wrote about what the Changed and Lasting Light looked like to them. The sights that are seen do repeat in exact detail this is true however take two people who saw from the same place and time. Both descriptions of the same event that they just witnessed will differ even if only slightly.

The Changed Light is very complicated and describing it 100% accurately is impossible. To start with in the exact same instant while seeing from within or from up past and over the upper right horizon one person could be looking to the right and the the other looking to the left for example. Each viewer may describe the exact same overall scene at this point but each viewer will see various different details in the area they were looking at. They will have no knowledge concerning sights they did not see but perhaps the other person viewing saw. I know that sounds weird but these are the types of effects that happen to the viewer. Complicated in many ways two people describing the same scene even at the same moment in time will have descriptions that vary from slightly to even greatly from each other. And then this effect continues to happen during the entire time that these two viewers see.

Descriptions of what happens in the Lasting Light differ. Now add in some of the many factors that also cause differences in the way ancient descriptions of the Lasting Light read to us today. For example the ancient writers spoke different languages, they were from many different cultures scattered all over the world and they were separated by time hundreds and thousands of years. Then today we read translated versions of their Return descriptions from many different ancient languages. No wonder why there is all sorts of confusion today. Plus throw in what might be the biggest reason for all the confusion and that is the fact that today the world is unaware of the existence of the Ancient Object and the Changed and Lasting Light event.

Basically no one knows what the ancient writers were talking about and yet many times today and throughout history we find reasons to persecute hate and even kill the other guy because he thinks about this differently. This is a tremendous ongoing tragedy on so many levels.

In the beginning there was the Awesome Good, The Ancient Man of Light created by a shadow that was cast from the surface the crossing ancient Object by a vertically standing crater rim piece. There are other human forms created by other shadows cast as well and to some degree I still think of this group of figures as the Guy and the Crowd the way I did as a young boy as I explain in various places but it does come back to the heart of the Changed and Lasting Light. And that main most important central element is the sight of the one and the only incredible ancient muscular Man of Light. The plain and simple fact is that soon the world will once again know that we all are in his image. We are all descended from this single incredible powerful muscular human form that is light. By default we are all created by nature equal. This recovered knowledge will create new discussion and should automatically settle and end old arguments.

As everyone knows the situation today in our world is a very unfortunate mess. Everyone agrees that only something big and profound can change the desperately dire situation the people of the world are in for the better.

It is my sincere hope that the situation between the people of the Earth will take a turn for the better and drastically improve once the crossing Ancient Object returns down and the Lasting Light is seen once again.

That's what has to happen if we have any change to survive into the distant future as a species.

The magnificent Ancient Man of Light was happening before Humans can into existence. The Man of Light will continue to be created by a shadow long after we are gone from the Earth. The question is how much longer will we be here to see his muscular back and his right side close up from above and behind within the misty white layer of clouds as he sits or crouches facing away into the distance of the Ancient Valley of Lasting Light? Will our distant descendants get to watch and see him as he then turns around to his left and stands up all in one fast motion to then go face to face with them as intended by nature? I hope so and I know that soon this will be your hope as well.

10 I'M ABSOLVED

I'm absolved, although it's not quite as simple as that however the fact is I am absolved.

Now many thousands of people have heard about the Ancient Object that I saw and the Lasting Light event that it causes having read from these pages. Even if people can't understand that my descriptions of the Ancient Object and the Lasting Light event are accurate every time another person hears, to me a small amount of the weight that I imagine is on me is removed.

In the big picture of things the only thing that I can do is provide accurate information and that's what I have done and that's what I continue to do. The weight that I imagine is all on me is actually on every individual inhabitant of our planet equally. I know this and eventually very soon everyone else is also going to know that this last statement is true.

The very intimidating, very Ancient Object Adoil is returning orbiting back home and when it intercepts and changes the light from the Moon and the background of space the Lasting Light Awesome Good will once again turn around and stand up after having been created, sculpted and shaped by the point of a fast moving obelisk shaped very tall black shadow that was cast by a vertically standing crater rim piece.

The speeding Ancient Object Adoil is the worrying part of all of this. The Moon's changed and intense Lasting Light formed after being slowed down and made solid looking by the Ancient Object's forces, then fantastically shaped and sculpted by the point of a shadow is the good news part of this that many people today mistakenly somehow seem to think they already understand. Because the speeding Object orbits, once again the very incredible very ancient Awesome Good will be created and then seen looking up, down at us, from down within, while rising with it's friends from over the Object's upper right horizon.

The amazing very incredibly muscular Awesome Good is returning home again as always!

A fantastic time in the history of all the people of the Earth is about to take place again!

I am absolved

ABOUT THE AUTHOR

I was born in Montreal Quebec, Canada in 1960. I believe that I probably saw the Object of the Crossing Down in 1972. At that time I was living in Lorne Park, Mississauga, Ontario. As a way for me to report and share my information for I started my returnviewersguide website in 2005. I am not a writer however for many reasons I knew it was time to attempt to describe what happens between the Earth and the Moon. My self edited writing is a difficult read. I know I have important information that I have to share and report by whatever means available.
Thank you to everyone who is able to overlook and see past my shortcomings as a writer.

Using a small backyard telescope I saw the Object cross down between the Moon and the Earth. I know that I have important information that needs to be reported and shared with people. I saw the Moon's transformed Changed and Lasting Light suddenly appear to the right of the crossing Object. I know that everything I saw repeats and happens again exactly the same way every time the Object crosses down between the Moon and the Earth. This means that when I describe what I saw I am also describing what will be seen in the future and also exactly what was always seen in the past by all of our ancestors. My descriptions are accurate, reliable and dependable although not professionally written. I know that because of the shear gravity of the situation I have to try to do something about it so that as many people as possible have a chance to see the very special Changed and Lasting Light from the Moon and the background of space. People need to know that it's not just the Object returning and crossing down but it's the very special sight of the Moon's Changed Light. That's important and this is why I have been forced by circumstance to become a writer.

My descriptions of the Object and the Moon's Changed Light are not things that I attempt to prove today because that would be impossible. Instead when the Object returns and changes the light from the Moon and then casts shadows into it's Changed and suddenly Lasting Light everything will become clear to everyone... Don Beeton

www.returnviewersguide.ca

www.ingramcontent.com/pod-product-compliance
Lightning Source LLC
Chambersburg PA
CBHW021412170526
45164CB00002B/609